**OTTMAR MERGENTHALER**

*The Man and his Machine*

Frontispiece: Ottmar Mergenthaler aged 25 years

# OTTMAR MERGENTHALER
# The Man and his Machine

*A biographical appreciation
of the inventor on his centennial*

## BASIL KAHAN

Introduction by Carl Schlesinger

Oak Knoll Press
New Castle, Delaware
2000

First Edition
Published by **Oak Knoll Press**
310 Delaware Street, New Castle DE 19720

ISBN: 1–58456–007–X

Title: Ottmar Mergenthaler: The Man and his Machine
Author: Basil Kahan
Typographer: Andrew Burrell
Publication Director: J. Lewis von Hoelle
Cover Design: Marius Kahan

Copyright: © 1999 Basil C. Kahan

**Library of Congress Cataloging-in-Publication Data**

Kahan, Basil.
 Ottmar Mergenthaler: the man and his machine: a biographical appreciation of the inventor on his centennial/Basil Kahan; with an introduction by Carl Schlesinger.
   p. cm.
 Includes bibliographical references (p. ) and index.
 ISBN: 1–58456–007–x
  1. Mergenthaler, Ottmar, 1854–1899. 2. Printers—United States—Biography. 3. Inventors—United States—Biography. 4. Linotype—History—19th century. I. Title.
 Z253.M483 B35 1999
 070.5'092—dc21
 [B]

99–045557

ALL RIGHTS RESERVED
No part of this book may be reproduced in any manner without the express written consent of the publisher, except in the case of brief excerpts in critical reviews and articles. All inquiries should be addressed to:
Oak Knoll Press, 310 Delaware Street, New Castle, DE 19720

Printed in the United States of America on 60# archival, acid-free paper.

# Contents

|   | | |
|---|---|---|
| | List of illustrations, diagrams, etc | ix |
| | Introduction | xi |
| | Comments and acknowledgements | xiii |
| | Prologue | 1 |
| 1 | **A watchmaker turns inventor** | 5 |
| | 1 Bucking the trend | 5 |
| | 2 Honing his skills | 10 |
| | 3 The land of opportunity | 11 |
| | 4 Opening the door to invention | 12 |
| | 5 Mergenthaler as the driving force | 18 |
| | 6 Portrait of an inventor | 23 |
| 2 | **The beginning of hot metal composition in newspapers** | 26 |
| | 1 Refining and redesigning | 26 |
| | 2 The syndicate moves in | 27 |
| | 3 In-fighting in the syndicate | 31 |
| | 4 Breakthrough – the prototype of the single matrix machine | 33 |
| | 5 The Mergenthaler Printing Company | 33 |
| | 6 The first Linotype is produced | 36 |
| | 7 The first machine goes to work | 40 |
| 3 | **Large scale production and ensuing contention** | 44 |
| | 1 Slipping schedules and teething troubles | 44 |
| | 2 Pressure to produce large numbers of Linotypes | 47 |
| | 3 Concerns about production and management | 49 |
| | 4 Reid's 1888 report and the start of strife | 51 |
| | 5 Mergenthaler's resignation | 53 |
| | 6 Production at Company factories after Mergenthaler resigned | 58 |
| | 7 Mergenthaler on his own again | 61 |
| | 8 Financing the new machine | 66 |
| | 9 Stilson Hutchins and selling the manufacturing rights abroad | 68 |
| | 10 Reid's 1889 report and resignation | 70 |
| | 11 Opinion | 73 |

| | | |
|---|---|---|
| 4 | **Hine takes over and Mergenthaler returns** | 74 |
| | 1 Mergenthaler reinstated | 74 |
| | 2 The British Linotype Company and the American connection | 77 |
| | 3 General progress and production at Baltimore and Brooklyn | 80 |
| | 4 The Typograph and the start of litigation | 87 |
| | 5 Financial considerations during Hine's term of office | 92 |
| | 6 Comment | 96 |
| 5 | **Dodge's rule and contention from other linecasters** | 97 |
| | 1 Developments in the Typograph case | 97 |
| | 2 A short digression about justification | 101 |
| | 3 The Monoline | 105 |
| | 4 Confrontation with Dodge | 109 |
| | 5 Mergenthaler and Clephane | 114 |
| 6 | **Mergenthaler** | 119 |
| | 1 The domestic picture | 119 |
| | 2 Relations with his men | 121 |
| | 3 Comments about his creativity | 123 |
| | 4 About his illness | 125 |
| | 5 Recognition during his lifetime | 132 |
| | 6 His death, his will and estate | 136 |
| | 7 Posthumous recognition | 138 |
| | 8 Mergenthaler – the man | 145 |
| 7 | **An overview of the British Linotype Company** | 151 |
| | 1 Background to the British launch | 151 |
| | 2 Preparing to launch the Linotype Company | 153 |
| | 3 Advertising the Linotype Company in the British Press | 156 |
| | 4 Legal considerations in setting up the Linotype Company | 163 |
| | 5 Public response to the launch of the Linotype Company | 165 |
| | 6 Early British reactions to the Blower Linotype | 169 |
| | 7 Progress of The Linotype Company | 172 |
| | 8 The Blower Linotype in the United Kingdom | 175 |
| 8 | **The Linotype – a technical summary** | 177 |
| | 1 Single operator linecasting machines | 177 |
| | 2 The individual matrix Blower Linotype | 179 |
| | 3 The Square Base Linotype | 183 |
| | 4 The Simplex Linotype | 185 |
| | 5 Significant enhancements to the Linotype | 186 |
| | 6 Constraints on materials | 188 |
| | 7 Maintenance and operation of the Linotype | 190 |

CONTENTS

| | | |
|---|---|---|
| 9 | **Facts and fancies** | 193 |
| | 1 Contemporary comments about the Blower Linotype | 193 |
| | 2 Contemporary reports of the impact of the Linotype | 195 |
| | 3 Reminiscences of experience with early Linotypes | 202 |
| | 4 Nailing the myths | 205 |
| | 5 The last word | 212 |
| 10 | **The Whittakers and the last Blower Linotype** | 213 |

**Appendices**

| | | |
|---|---|---|
| A | **Selected Chronology** | 219 |
| B | **Glossary** | 224 |
| C | **Biographical notes** | 227 |

**Bibliography** 232

**Index** 239

# *Illustrations*

| | | |
|---|---|---|
| 1 | Mergenthaler's birthplace. The schoolhouse at Hachtel before refurbishment | 7 |
| 2 | Portion of letter written to Herr Roller | 8 |
| 3 | The movement of the village clock at Ensingen | 9 |
| 4 | Mergenthaler and friend dressed for a party | 13 |
| 5 | Fragment of output from the transfer machine | 15 |
| 6 | Diagram of second band machine including detail of band (on left) | 20 |
| 7 | An artist's impression of the prototype of the Blower Linotype | 34 |
| 8 | Illustration from the first article about the Blower Linotype | 42 |
| 9 | A lampoon of Whitelaw Reid | 55 |
| 10 | Mergenthaler's letterhead and a portion of map showing his works | 63 |
| 11 | The 'Improved Linotype' known as the Square Base | 67 |
| 12 | Part of the *New York Tribune* Sunday supplement, 19 May 1889 | 78 |
| 13 | Woodcut from the *Louisville Journal* to which Mergenthaler took exception | 80 |
| 14 | Letter headings used by the Mergenthaler Printing Company | 83 |
| 15 | The Rogers Typograph | 88 |
| 16 | The first American Linotype advertisement | 89 |
| 17 | Line of matrices with double wedge space bands | 101 |
| 18 | Mergenthaler's graduated wedge justifier | 102 |
| 19 | Copy of Mergenthaler's diary for 21 January 1893 | 103 |
| 20 | Copy of Mergenthaler's diary for 18 October 1893 | 104 |
| 21 | The Monoline | 106 |
| 22 | Mergenthaler Linotype Company letterhead and picture of the Brooklyn factory | 111 |
| 23 | The Mergenthaler House at 159 Lanvale Street West | 126 |
| 24 | Plaque by the front door of the Mergenthaler House | 127 |

| | | |
|---|---|---|
| 25 | Fragment of a letter from Mergenthaler to an acquaintance | 128 |
| 26 | House at Deming, NM, destroyed by prairie fire in 1897 | 130 |
| 27 | Cover from the Catalogue of Ott. Mergenthaler and Co | 133 |
| 28 | Linotype Company advertisement showing Simplex (model 1) Linotype with Antwerp Gold medal | 134 |
| 29 | Emma and Pauline at a ceremony honouring Mergenthaler | 139 |
| 30 | The school house at Hachtel refurbished (c 1925) | 140 |
| 31 | Mergenthaler Hall at the Johns Hopkins University | 142 |
| 32 | Memorial plaque to Ottmar Mergenthaler at the Johns Hopkins University, Baltimore | 143 |
| 33 | The school house at Hachtel refurbished (c 1954) | 144 |
| 34 | Mergenthaler postage stamps | 146 |
| 35 | Entrance to the Mergenthaler Vocational Technical High School | 148 |
| 36 | A woodcut of the Blower Linotype in the UK | 155 |
| 37 | A Linotype Syndicate advertisement | 158 |
| 38 | Scheme of negotiations for the British rights in the Linotype | 163 |
| 39 | A lampoon of John Charles Cottam | 167 |
| 40 | The Thorne typesetter | 171 |
| 41 | Battery of Blower Linotypes at the *Sheffield Telegraph* | 175 |
| 42 | Cycle of operations for producing lines of type | 178 |
| 43 | The Blower Linotype with details | 181 |
| 44 | The distributor bar and matrices | 182 |
| 45 | Assembler box and spaced line of double-letter matrices | 187 |
| 46 | Composing room at the *Sheffield Telegraph* (c 1892) | 198 |
| 47 | Composing room at the *Sheffield Telegraph* (c 1901) | 199 |
| 48 | Miller and Letsch 50 years apart | 203 |
| 49 | A piece of fiction concocted by the Linotype Company | 208 |
| 50 | Another lampoon of Whitelaw Reid | 210 |
| 51 | Peter Whittaker and Stan Nelson with the Blower Linotype | 215 |
| 52 | Pastiche advertisement of the Blower Linotype | 216 |

# *Introduction*

Basil Kahan's book on Ottmar Mergenthaler could not have come at a better time. As I write this the signs are positive that America and England are showing increasing interest in preserving and examining our industrial and economic past. In both countries small home-grown museums and giant collections are being organized, sparked by some desire deep within our psyche to hold on to the Industrial Revolution, appreciate and preserve our past and know our artisan's roots.

In some of these converted home garages, barns, old factories and warehouses, Linotype machines are being restored and returned to a working life. Young people will view these antiquated mechanisms as objects of great curiosity, for they will have no memory of our vaunted mechanical past. Today's infants come out of the womb, it seems, with a computer mouse in their hand, ready to surf the net. We hope their search leads them to this book. It is not only yesterday preserved – it is 19th century business history, with some of the sharp practices of those days still used in today's world.

The anniversary of an event rekindles interest in the original occurrence. In our special field of remembrance, we should ponder October 28, 1999, the 100th anniversary of the death of Ottmar Mergenthaler. At age 45, one of the world's greatest inventors died too young, and of a disease – tuberculosis – brought on by many human and industrial factors. Among them was the great stress put upon Mergenthaler by his business associates, many of them eager to exploit the inventor and his miracle machine to make a quick buck.

Eager to please his backers, yet working alone against almost impossible deadlines, Mergenthaler could not build the machine he envisaged until he first invented and designed the special tools he needed. To report progress he had to burn the midnight oil night after night. It is easy to visualise the effect these conditions had on his physical and emotional health.

Fate put this basically honest mechanical wizard, a workaholic who thought and acted more like a blue-collar factory manager than a business tycoon, into the same pot with the wealthy patrician Whitelaw Reid, publisher of the *New York Tribune,* and Philip Dodge, patent attorney, inventor and president of the Mergenthaler Printing Company throughout the 1890's. Mergenthaler's disagreements with both men and his extreme dislike for them ate at his insides. As Kahan so ably documents, the inventor was certain they were trying to cheat him out of his royalties, degrade his efforts on behalf of the company and make him the scapegoat of misguided company policies.

Add to this mix the chameleon Stilson Hutchins, publisher of the *Washington Post* and champion of the Linotype's great potential. Be sure to read Kahan's characterization and summary of Hutchins's personality and career in the book's Biographical Notes. Hutchins's deviousness, combined with his ability to foresee the Linotype's great future, added to the misadventures of the Linotype Company, especially during the establishment of the British syndicate which planned to build machines in the United Kingdom.

In that caper, the author shows how Hutchins tried to flog 60 of the outmoded Blower Linotypes to the British, all the while keeping secret from them that in America the newer, considerably improved model was the latest machine in production and manufacturing of the Blowers had ceased. Hutchins, who was empowered by the company to sell the English and Continental manufacturing rights, stood to make a lot of money in commission if he could pull the deal off. He did.

If all the above sounds too dramatic, Basil Kahan convincingly proves that once again, truth is stranger than fiction. Twists of fate, conflicting personalities, shady business tactics, the shredding of an ethical fabric . . . all are found here in fascinating detail.

Dr. Kahan is a convincing writer. He has done eight years of painstaking research in the United States, England and Germany. He certainly knows his subject. I wonder how this story would be received if it had been done in the form of a stage play or movie. The excitement and drama is there. Furthermore, the undeniable contribution to graphic arts history is also there. We know more about Mergenthaler – a three-dimensional Mergenthaler, with good points and bad – than ever before. That's a good thing. His invention rivaled Gutenberg's. It opened the floodgates of literacy to millions.

Faster typesetting lowered composition costs and produced more and cheaper printing in most of the world's languages. Since printing preserves people's thoughts by converting them to a permanent form, through this process we can know much more about our past. If we thank Gutenberg and Mergenthaler for their major role in helping to make that possible, we should give a deep tip of our hat to Basil Kahan as well, for his zealous research and major effort in writing this important book.

Carl Schlesinger

June 1999
*Rutherford, New Jersey*

# *Comments and Acknowledgements*

Although I am not a printer there is printer's ink in my veins. My maternal grandfather, Charles French, born in 1860, was a compositor who went to work in France. When composing machines were introduced he became a 'pica-thumping' proof reader and retired as head proof reader on the Paris edition of the *New York Herald-Tribune* (now the *International Herald-Tribune*). My mother Edith, a proof reader, met my father Waldo Kahan, a Linotype operator, in the proof room of that paper. He had gone to the *Herald* for keyboard training. My father set up a trade typesetting business in London shortly after the end of the second world war. During a slack period while I was waiting to go to university, my mother and I erected an English Model 3 Linotype that we had acquired in pieces. I also worked occasionally as a Linotype mechanic for my father. For some 30 years I had little connection with printing. I taught Mathematics to Engineering and Science students for over twelve years and then joined a major computer company and specialised in Statistics and Operational Research. My interests turned full circle when I became involved with text, graphics and desktop publishing. After taking early retirement I went back to university to investigate the early history of the Linotype in the United Kingdom. In the course of my studies I found that many widely accepted accounts of Mergenthaler and his invention were spurious so I decided to try to sort fact from fiction in a new account of Mergenthaler's life and the early days of the Linotype. This is particularly appropriate on the centennial of his death which occurs so soon after his machines have been displaced by the very technology that I have used to write this book.

Mergenthaler was an impecunious German teenager who emigrated to the USA and invented the Linotype. Two conflicting opinions attributed to Thomas Alva Edison described the machine as the eighth wonder of the world and one of the ten outstanding inventions of the nineteenth century.

This biographical appraisal of Mergenthaler is intended for the general reader so the footnotes usually given in a researched account have been omitted deliberately. However, there is an extensive bibliography at the end of the book for readers who want more detail. This book is intended to amplify the account given in Carl Schlesinger's 1989 edition of 'The Biography of Ottmar Mergenthaler – Inventor of the Linotype' which was written by Otto Schoenrich at the dictation of the terminally ill inventor. Although an autobiography in all but name it will be referred to as 'the biography' throughout this text. Autobiographies are usually biased; they present the author's point of view and, as Mergenthaler admitted at the end of a letter to an unidentified friend, he intended his 'history' to cause a sensation, which makes it important to investigate all shades of opinion.

In retrospect the development of an effective Linotype was remarkably quick, especially when one considers the time taken to design and produce a new plane or car at the end of the twentieth century, with the support of computers and cad/cam techniques. Those were days of exploration, when people built machines and modified them if they broke down. When the work that led to the Linotype began no one had realised the metallurgical problems associated with effective type metals for composing and casting machines, nor the requirements of an alloy from which to make matrices. Such problems were tackled 'in the field', rather than in the laboratory. At that time users did not realise how quickly mechanical systems deteriorate if not regularly maintained.

Several themes run through this book besides the life of Mergenthaler, his inventions and his associates. There was manoeuvring between various groups of sponsors, problems of cash flow in the early days, litigation, and the negotiations to set up the British Company. I have tried to cross-check all sources of information in the hope of giving an honest word-portrait of the man, 'warts and all'. Over some issues Mergenthaler was appallingly treated, particularly by some of the senior executives in the company that bore his name, but he revealed many negative traits in attempting to justify his actions. To get a fresh opinion, I lent the biography to a German friend who said as she returned the book, 'Oh that poor man, but he really did bring much of his troubles on himself.'

In the course of my research I have noted a number of anomalies; some concerning dates that have obviously been misremembered over the years and others about the spelling of names, particularly of Germans. When in doubt I have followed the style used in the biography. A later chapter is devoted to some of the more outrageous myths about Mergenthaler that have circulated, apparently unchallenged, for the best part of a century.

I have been helped by so many people over the years that I hope that in naming some I will not offend others who have unintentionally been overlooked. My particular thanks go to Carl Schlesinger and Professor Corban Goble who have gone out of their way to encourage my research and to my sons, Gez, who read the manuscript and made many helpful suggestions and Marius, who produced the original design of the dust jacket. Others include: Professor Michael Twyman, former Head of the Department of Typography & Graphic Communication at Reading University and his staff, and the staff of the Reading University Library.

From the USA I am indebted to my cousins Monica and Bob Harris for their hospitality and particularly to Monica for chasing up references in the Library of Congress when I was not Stateside. Others who went out of their way to help me include: Ottmar Mergenthaler's granddaughter, Miss Nancy Perkins; Mr Ronald Mergenthaler, grandson of Ottmar's half-brother Fritz; Dr Elizabeth Harris, formerly Curator of the Division of Graphic Arts at the Smithsonian Institution, Washington, DC; Stan Nelson, Specialist at the Smithsonian, who explained details of the second band machine and demonstrated the Blower Linotype; Richard Huss; Gene Lawrence, Principal of the Mergenthaler Vocational Technical High

## COMMENTS AND ACKNOWLEDGEMENTS

School; Edward Leon of CHAP (Cultural Heritage and Architectural Preservation), Baltimore; Joan Gratton of Johns Hopkins University, Baltimore; Joe Mazzella, formerly of Mergenthaler Hot Metal; and Mark Barbour, Curator of the Museum of Graphic Communication in Buena Park, California.

From the St Bride Printing Library in London I particularly want to thank James Mosley, Nigel Roach and Lynn Arlotte for their guidance through the 19th century trade papers. Other libraries that helped me included: the Guildhall Library; the British Library; the Newspaper section of the British library at Colindale; the Science Library and the Patent Office Library in Chancery Lane; The Public Records Office at Kew; the London College of Printing Library and The Library of Congress in Washington, DC.

My sincere thanks go to the Whittaker family for letting me examine the late Peter Whittaker's archives at his private linecasting museum (now sited at the Museum of Science and Industry in Manchester). Chris Hall and Mrs Christine Gale, formerly of M. H. Whittaker & Son (Holdings) Ltd, coordinated my visits and gave me photocopies of rare documents.

I would also like to thank Mr John F. Dixon, formerly Company Secretary, and Mr Martin Boothman, formerly Managing Director of the British Linotype-Hell Company for their assistance and encouragement.

I am indebted to Heidelberger Druckmaschinen AG for permission to reproduce illustrations from publications of the American, British and German Linotype Companies.

The choice of illustrations is always a problem, particularly when some are 100 years old. Photographs fade, newsprint crumbles and even good microfilm copies may be distorted. One has to decide whether to use a poor copy to make a point or to omit it because of its low quality. Where I thought that an illustration would improve the message I have sought the best available. I hope that readers, and particularly reviewers, will take this into consideration when appraising this book.

Finally, any errors in this book are my sole responsibility.

**OTTMAR MERGENTHALER**

*The Man and his Machine*

# *Prologue*

In July 1886, Ottmar Mergenthaler, the principal inventor of the Linotype, installed the first production model at the *New York Tribune* and started the hot metal era of newspaper production that lasted for almost a century.

From the fifteenth century, when Gutenberg introduced printing with movable metal type, the trade had been dominated by letterpress. A hand compositor set text one character at a time by picking pieces of type from a case and assembling them in a hand-held frame called a 'composing stick'. Each line had to be the same length so that the type could be 'locked up' for printing. Within each line the space between each word had to be equal – this was called 'justification'.

Printing had been a totally man-powered craft until 29 November 1814, when John Walters, in London, England, had used Koenig and Bauer's press to print *The Times* 'by steam'. The resulting increase in printing speed pinpointed typesetting by hand as the major bottleneck in newspaper production and started a long-term quest for an effective typesetting machine. Nineteenth century printers would have expected an acceptable composing machine to perform three tasks: to set type; to justify that type; and to distribute the type at the end of the job.

In 1822, Dr Church, an American inventor living in England, obtained the first patent for a typesetting machine that was never built. At the time his machine was described as a hoax, because the thought of setting type mechanically was as fanciful as a flying machine! How could this man, with no background in printing, design such a device? Nevertheless, some of Church's ideas were adopted by other inventors. To increase the speed of composition he designed a keyboard so that the operator could set type with both hands. He also suggested stacking type in a vertical magazine with a guide plate to keep it facing in the right direction as it was assembled. Finally, he avoided the need for distributing type at the end of a job by proposing that type be recast after use, which meant that printers would always work with new type.

Church, and many others, tried to emulate the hand compositor using movable type, but at least two operators were required to make those early machines effective – one to set text continuously and another to justify it. An operator who broke off setting to justify lines took more than twice as long as two people working together. Most of these inventions were not successful commercially and it was reported that the cellars at *The Times*, which tried to encourage new inventions, were blocked with discarded machines. The bottleneck in the composing room became so critical that in 1869 the *New York World*, then owned by Manton Marble, proposed that publishers subscribe to a prize fund of $500,000 for a machine that would speed up composition and reduce costs by at least 25 per cent. No records have been found to confirm either that the money was raised or that any prize was awarded.

However, inventors did not restrict themselves to letterpress printing with movable type. They investigated a number of alternative processes, including newer technologies such as lithography and photography, which were not yet sufficiently advanced to be applied commercially.

Others tried to mechanise business correspondence which led to the invention of typewriters. These devices, intended for personal use rather than the mass production of printed matter, have the advantage that there is no type to be distributed but the drawback is that they can only print the characters supplied with the machine.

James Ogilvie Clephane (see biographical notes) the leading court reporter of the day, who covered all the important trials in Washington, DC, was very dissatisfied with hand-written court reports that took too long to prepare and sought a more effective method of producing them. His early interest in typewriters started the line of research that led to the invention of the Linotype. In 1866 he heard about a typewriter invented by a Mr Pratt of Selma, Alabama, but after investigation decided that it was not of 'a practicable character'.

The next year Messrs Densmore and Sholes, who were developing a typewriting machine, contacted Clephane in Washington. He agreed to pay Sholes $150 for every machine 'that he might build by hand for him to test in a practical way'. Clephane tested five or six models to destruction, each being superior to the previous version. The experiments lasted until 1873 when the armaments manufacturer, E. Remington & Sons of Ilion, NY, started to produce the Sholes machine commercially.

In late 1874 Clephane gave the typewriter a thorough test when he reported a celebrated burglary trial that lasted six weeks. He used seven typewriters of the first batch, with seven typists and seven stenographers, to transcribe his notes as they were taken in court. Those early users were rather slow and unskilled but, by using duplicating paper, he was able to distribute three copies of the revised, stitched and indexed transcript to counsel by 8 o'clock each evening. Clephane was offered a substantial investment in the typewriter, which would have made him a wealthy man, but he was not completely satisfied with the output and abandoned typewriters to investigate other methods for producing court reports more quickly.

Clephane and his associates tackled the problem in several ways, none of which used movable type. It is not widely known that in 1875, **before** he had any contact with Ottmar Mergenthaler, Clephane had tried to cast a line of type from a mould made by using a Remington typewriter to impress characters into soft plastic material.

Clephane thought that a wheel machine similar to the Phelps Printing Telegraph used by a Mr Royce at the Western Union Telegraph offices would be the ideal device. After Clephane had discussed the matter with him, Royce wrote to Mr Phelps, the inventor, who thought that it would be possible to construct a machine to print on a page. However, Phelps would not undertake the project because anything that he did on the premises would belong to Western Union and he felt that he was too old to enter a comparatively new field of invention. On receiving this reply, Royce told Clephane that he had seen a model of a printing telegraph that seemed to be 'somewhat in the direction of such a machine as he was looking for'.

The inventor, Charles T. Moore of West Virginia, came to Washington at Clephane's expense. Although relatively young and with little practical experience, Moore clearly 'possessed genius' and was confident that he could do what he claimed. Clephane and others financed him for about a year, paying for the preparation of drawings, the construction of models and for expenses in New York, Philadelphia and Washington. This first contact with Moore involved considerable outlay but did not result in a practicable machine. In July 1875 he submitted a new design that was dropped in August 1875, because no one was willing to risk more money on the venture.

From then until early 1876, Clephane and his associates sponsored a Mr George H. Morgan who was developing a new style of typewriter. During this time Clephane paid large sums to build models of inventions and to apply for patents in the USA and other countries. He abandoned this venture when all Morgan's efforts were unsuccessful.

In the meantime, Moore, who continued to claim that he could produce a satisfactory device based on his plans of the previous summer, obtained an advance of $1,500. This device was intended to produce a printed strip of special paper to be cut and pasted into pages of justified text which would be transferred to a stone to make a master for lithographic printing. This invention, known as the transfer machine, was built by a Mr Grant of Baltimore but on 3 July 1876, when Clephane went for a demonstration, it would not print. Moore assured him that it would work properly on 6 July; however when Clephane arrived 'he found Mr Moore and Mr Grant engaged in a very earnest altercation'. The engineer claimed that the design was faulty and Moore blamed the engineer . . .

. . . but if that machine **had** worked

this book would not have been written . . .

# ONE

## *A watchmaker turns inventor*

In 1872 the United States of America was an exciting new country that barely a century before had been a colony and had only recently emerged from the trauma of civil war. A country whose example had inspired the French to propose a statue of Liberty in the previous year. A country that was attracting people from all over Europe as the land of opportunity, free from feudal government and, hopefully, with none of the racial, religious and political intolerance that plagued their homelands. A land that welcomed vigorous, intelligent, hard working individuals with the pioneering spirit and self-confidence to give as much to the country as they gained from it. A land where most immigrants did not speak the official language.

So what was going through the mind of the impecunious teenager who disembarked from the SS *Berlin* at Locust Point, Baltimore on 26 October as he started out for Washington, DC. The prospect was both encouraging and daunting but at least he had a place to go. He was of medium height, sturdy and healthy, a skilled watch and clock maker, an outstanding craftsman who could turn his hand to anything in light engineering. He could face the future without fear and may have been wondering about the quirks of fate that had taken him from his home in the province of Württemberg in Southern Germany, to a strange country, with a strange language, on the other side of the Atlantic ocean. Maybe he was reflecting on the past as he contemplated the future.

### 1 *Bucking the trend*

Comparatively little is known about Ottmar Mergenthaler's early life. He gave few details in *The Biography of Ottmar Mergenthaler – Inventor of the Linotype* (cited elsewhere as 'the biography') that he dictated to Otto Schoenrich when he was dying. The family had been peasant farmers for generations until his father, Johann Georg Mergenthaler (1820–93), broke with tradition and became a school teacher. He married Rosina Ackerman (1828–59), who came from a family of teachers. In the biography Ottmar was described as being the third in a family of five children, but in his two marriages Johann Mergenthaler actually fathered ten children, only five of whom reached maturity. Ottmar was born on 11 May 1854 in the schoolhouse at Hachtel, a hamlet a few miles south of Bad Mergentheim, in the German province of Württemberg, see figure 1. The date was erroneously printed as 10 May in the biography. [In his PhD dissertation Professor Corban Goble noted that Mergenthaler had amended the date in pencil in the copy that had been specially bound for Emma, his wife.] Johann Mergenthaler was transferred to Neuhengstett in the autumn of that year and four years later moved, at his own request, to Ensingen, where Ottmar grew up. He was five years old when his mother died of childbirth fever after bearing seven children in her ten-year marriage. The family was looked after by an aunt, Wilhelmine Ackerman, until Johann Mergenthaler married Caroline Hahl in February 1861.

Ottmar received a primary education at his father's school. His lessons included music which he enjoyed throughout his life. He was also very neat and precise as shown in figure 2 by the portion of a letter written to his godfather Herr Roller on April 12, 1868. Very faint guide lines were scored in the original. The family was so poor that in years when the crops failed they could not even afford to buy potatoes and the children often went to bed hungry. Everyone had to help with the household chores. Ottmar helped to prepare meals, wash dishes, make fires in winter, till the garden in summer and look after their domestic animals. He also had to help keep the church tidy. He said that it was all work and no play but he accepted the position willingly, because he was used to it. Obviously he had some spare time, and he showed such mechanical flair in his hobbies that the other children in the family called him *Pfiffikusmärle*, an old German expression for a 'whiz kid'. He may have inherited this aptitude from his maternal grandfather, Johann Christian Ackerman (1791–1853), who was a surveyor and architect as well as being a school teacher at Erbstetten. Ottmar carved models of animals with his penknife and made animal shaped wooden biscuit moulds that looked so professional that people wondered that a young boy could produce such work. In later life he recalled how he had seen the need for an improved chuck for lathes when still a boy in Germany. After working on it diligently he showed it

Geburtshaus Mergenthalers in Hachtel

Geburtshaus mit den Fenstern des Geburtszimmers

Figure 1   Mergenthaler's birthplace: the schoolhouse before refurbishment
*Source: Otto Schlotke, Ottmar Mergenthaler's Jugendjahre, pp 4 and 5*

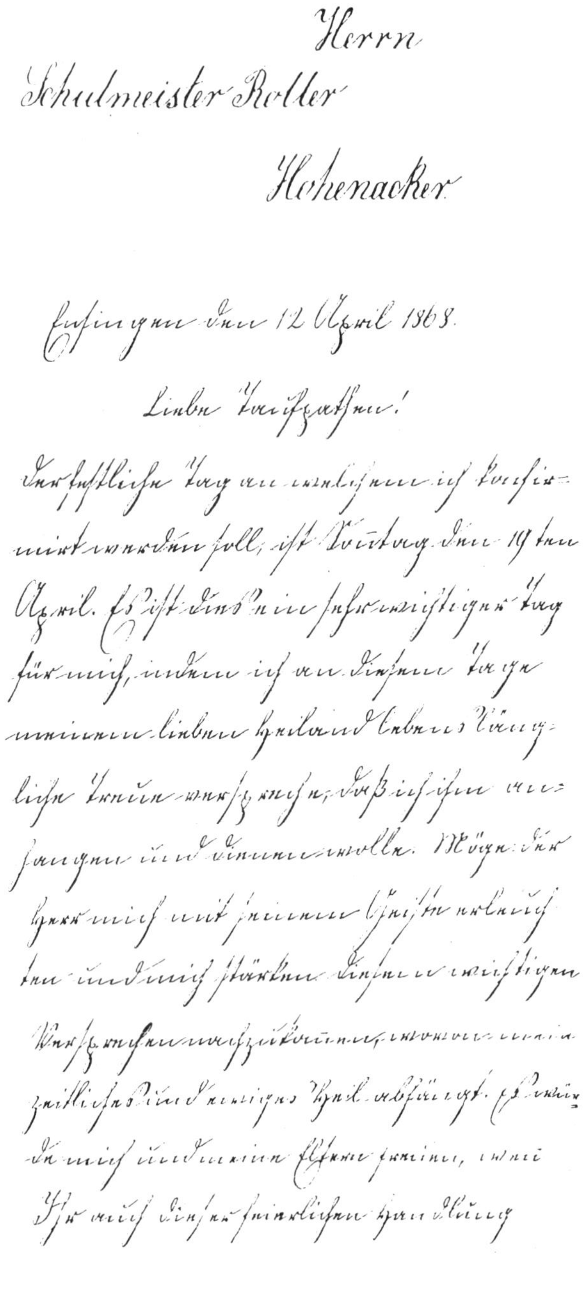

Figure 2 Portion of a lettter written to Herr Roller
*Source: Mergenthaler's family papers*

to a neighbour, a machinist. He was very hurt when the man gave it a casual glance and without saying a word reached into his bench and brought out an almost identical device.

Ottmar had a deep interest in clocks and kept several in working order. His most famous boyhood achievement was secretly restoring the village clock at Ensingen which clockmakers from Stuttgart had declared to be beyond repair. The clock tower was dangerous and strictly 'out of bounds' and he was punished for entering it without permission (which would have been refused) but, having made the repair, his father reluctantly allowed him to maintain the clock. That historic clock is now installed in the old Hachtel school house. The movement is shown in figure 3.

Johann Mergenthaler, an intellectual snob who was motivated more by status within the community than by salary, wanted his children to follow in his footsteps. He was obviously disappointed when in 1868 at the age of fourteen Ottmar refused to follow his elder brothers into teacher training. When his father told him that the matter had been decided and that there was to be no argument, the lad retorted that if the profession was so good, why did his father complain about the poor pay, the lack of prospects for advancement and

Figure 3  The movement of the village clock at Ensingen
*Source: Otto Schlotke, Ottmar Mergenthaler's Jugendjahre, p 11*

harassment by the Inspectors of Schools. This refusal showed considerable courage; few boys would have defied a German schoolmaster at that time. Ottmar wanted to be an engineer and cited his aptitude with tools and the repair and maintenance of the village clock as reasons for wanting to work with machinery; but he had only a primary education and the family could not afford the secondary education that was a prerequisite for an engineering course. When the boy threatened to take a job as a mechanic, a galling prospect for his father, the elder Mergenthaler spoke to several craftsmen in the district, but all said that there was little demand for their skills. It is evident that Ottmar respected his father and regretted this disagreement about his future – he desperately wanted to succeed and make his father proud of him. Finally it was suggested, presumably by his stepmother, that he should be apprenticed to her brother Louis Hahl, a watch and clock maker in Bietigheim. The terms of his apprenticeship were that he was to serve four years without wages, pay a small premium, furnish all his own tools, and receive board and lodging from his stepmother's brother.

## 2 Honing his skills

Mergenthaler's four-year apprenticeship passed quickly. The hours were long but the atmosphere was pleasant and he got on well with his fellows. He was a conscientious worker and attended optional technical courses in the evenings and at weekends to improve his qualifications. Some years later he used that knowledge in the preparation of technical drawings and the drafting of patent applications. He became so competent that, although his articles did not require it, he was paid a journeyman's wage during his final year's training; the only time in 30 years that Hahl had so rewarded an apprentice. Mergenthaler claimed: 'Above all, watchmaking taught me precision.' He learned how to temper springs, cut out the finest teeth, make pins, bore jewels with firm, steady pressure and understand very complicated movements. He saw that if a watch is to be accurate, its mechanism must be considered as a whole. Every new addition must harmonise with all the other parts to form a unit which is at once refined and intricate.

Prospects were not promising at the end of his training. Every vacancy was being filled by soldiers returning from the Franco-Prussian war; furthermore, Ottmar was due to be drafted into the army, where his elder brothers were already serving. The people of Southern Germany objected strongly to the yoke of Prussian militarism so, along with thousands of others, he decided to emigrate to the United States. Louis Hahl's son August, had established an instrument workshop in Washington, DC, which made electrical clocks and bells, and instruments for the United States Signals Service. Ottmar applied to him for a job and Hahl promptly advanced money for his passage, to be repaid out of his wages.

## 3 *The land of opportunity*

Mergenthaler travelled steerage from Bremen on the SS *Berlin*, one of five hundred passengers. At this time he has been described as a lithe, comely fellow of eighteen, about five feet seven inches tall, with a large well-shaped head on broad shoulders. He had a silver pocket watch that he adjusted daily so that at the end of the voyage it was as accurate as the ship's chronometer. He brought a round-topped wooden trunk full of clothes and $30 in paper money. On arrival in the United States Ottmar went straight to the Hahl shop in Washington.

He started to work for his step-cousin immediately. Although electrical instruments were new to him, he soon became as efficient as any of his fellow-workmen. Within two years he took a leading place in the shop, acting as foreman and, when Mr Hahl was absent, as business manager. Mergenthaler's routine work included making meteorological instruments such as heliographs, registers for wind velocities, and rain and snow gauges for the United States Signal Service. He was also responsible for most of the experimental work carried out at the Hahl shop and for the final designs of many standard devices. At that time Washington was a major centre for inventions made both in the United States and abroad and a model had to be submitted with each patent application. Mergenthaler built such models; he had a flair for transforming an inventor's concepts into working machinery. This brought him into regular contact with inventors and long before his twenty-first birthday he had applied his expertise to so many machines and instruments that he was generally recognised as an outstanding craftsman. He liked this work, which was to have a great impact on his future, because, naturally, it stimulated his own creativity. It is unlikely that he would have been so inspired if he had remained in his native Germany.

Mergenthaler tried to integrate into the American way of life as soon as he arrived in the USA. Speaking only German when he landed, he quickly became fluent in English, but it was said that he never completely lost his guttural accent. He used English in all his business correspondence and diary entries, but never mastered English spelling and tended to take any statement in English much too literally. He became an American citizen in 1878 and considered himself to be American. However, he had a circle of German friends with whom he spent much of his spare time. In his short biography of Mergenthaler in *Leading American Inventors*, George Iles reported the comments of Mr Henry Thomas, a close personal friend of the inventor: 'Those formative years in Washington, Mergenthaler was wont to regard as the happiest of his life. He was one of a coterie of young Germans who lived together, sang together, and often took long walks together. Early on Sundays we were wont to stroll to Great Falls or Chain Bridge, halting at the farmhouse of a German friend. At his hospitable board we refreshed ourselves with clabber, potatoes in uniform [jacket potatoes with sour cream], black bread, and beer in moderation. Ottmar, reserved and almost silent with strangers, always let himself go in our company. He was a generous comrade, complying and kind, no spoil-sport. His voice, a fine barytone, was often heard in a repertory of German songs and ballads. In those days his health was vigorous and his step elastic. He gave the impression of being hale and hearty at fourscore. We were all ambitious, but he brought it further than any one of us all.'

Towards the end of 1873 Hahl's business came under pressure due to financial panic throughout the USA. Although Hahl lost work and was forced to lay off some employees he retained Mergenthaler. In January 1874, despite Mergenthaler's reservations, Hahl moved his business to Baltimore, a more important industrial centre with lower overheads. He opened a small shop at 13 Mercer Street, three doors from the premises of printer Charles W. Schneidereith. Although trade did not improve the workshop kept in contact with inventors.

Mergenthaler apparently retained his sense of fun even when under stress at work. Figure 4 shows him with his friend Mr Thoss dressed for a party in about 1880, when the Hahl company was in severe financial straits.

## 4 *Opening the door to invention*

In August 1876 Charles Moore consulted August Hahl about the transfer machine mentioned in the Prologue. He named James O. Clephane, his brother Lewis Clephane, Maurice Pechin and J. H. Crossman, all of Washington, DC, as sponsors. Moore attributed the failure of his machine to defective

Figure 4  Mergenthaler and friend dressed for a party
*Source: Baltimore Sun, 13 November 1960*

workmanship by the engineer Grant. Mergenthaler's opinion was that the model was well made but the design was flawed. He said that he could redesign the device to overcome these drawbacks. Mergenthaler and James Clephane established an immediate rapport and they became firm friends. L. Clephane and J. H. Crossman then agreed to pay August Hahl $1,500 if he produced a workable machine. In January 1877, Hahl and Mergenthaler made an eight-letter instrument that showed the principle of the rotary machine and they were paid the promised amount.

This was Mergenthaler's introduction to printing and he threw himself whole-heartedly into the project. Instead of the archetypal impecunious inventor seeking a sponsor, there was Clephane the enthusiast, full of ideas, able to raise capital, and seeking an inventor to turn those ideas into effective machinery. Mergenthaler, who did not often give credit to others, freely acknowledged Clephane as the originator of the line of research that led to the invention of the Linotype. He may have been rather too lavish in his praise because, at that time, many inventors had similar aims.

The lines of research proposed by Clephane always differed radically from those of inventors of machines based on emulating a compositor using movable type. Although he did not continue experimenting with typewriters, all Clephane's machines were based on the idea of making impressions; his concept of machine composition did not call for the distribution of type, because typewriters produce print immediately from an inexhaustible, but limited, master alphabet. Further, every machine was designed to be operated by just one person.

Once the pilot machine had been accepted Clephane contacted friends to raise further funds to pay for a full-sized machine. As no one else would contribute, Clephane supplied $1,000 to build the machine, provided it was designed to produce text in a continuous printed strip, even though his collaborators had urged him to develop a machine to produce output by the page. Mergenthaler constructed the full-sized machine, made his own drawings and did most of the machine work. It was completed in August 1877,

> . . . printing on an endless narrow strip, [it] worked very rapidly; mixed composition could be done with great ease by means of a single shifting of the type wheel carrying Roman and Italic faces; the spacing of the characters seemed to be very accurate, and the print was sharp and clear. Everybody was delighted.

Unfortunately, the impression did not always transfer properly to the lithographic stone; Mergenthaler and Clephane decided that there were too many stages in the process to establish proper control, so it was decided to change the system. Although further research on the lithographic process was abandoned due to erratic output, it did occasionally produce passable results. Reporting offices were opened in Washington, Chicago and New York and all did considerable commercial work.

It is only fair to point out that printers would have found the quality of this process far from satisfactory, but engineers and shorthand reporters who wanted a quick turn round of legible text, regardless of style, may have found it acceptable in the short term. Although Mergenthaler had shown the process to be feasible he had also shown the need for much development work before the trade would accept it. Output from an early version of the transfer machine, see figure 5, shows clearly that despite Mergenthaler's skill as a craftsman, he did not appreciate the requirements of high quality printing.

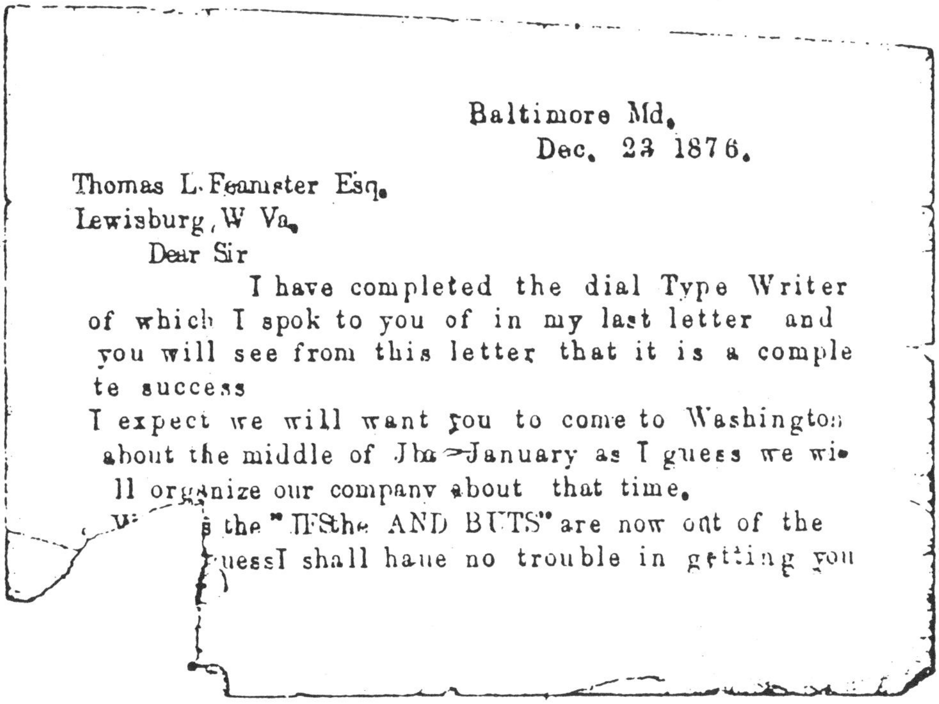

```
                              Baltimore Md,
                                Dec, 23 1876,
Thomas L. Feamster Esq,
Lewisburg, W Va,
       Dear Sir
              I have completed the dial Type Writer
       of which I spok to you of in my last letter  and
       you will see from this letter that it is a comple
       te success
       I expect we will want you to come to Washington
       about the middle of Jba →January as I guess we wi•
       ll organize our company about that time,
             v:    s the " IFSthe AND BUTS" are now out of the
                   uessI shall haue no trouble in getting you
```

Figure 5  Fragment of output from the transfer machine. Copied from a photocopy of the original, the location of which has not been traced

At this time the National Machine Printing Company was organised under the laws of the District of Columbia with capital of $28,000. Lewis Clephane was elected President of the company and Messrs J. O. Clephane, Moore, Royce, and Pechin assigned all their existing and future interests in typewriting and typesetting machines to the company. As mentioned in the prologue, Clephane had tried to cast type in lines in 1875. When the first Mergenthaler machine was completed Clephane tried to use it to indent tin-foil, hoping to take an electrotype cast from it. Later trials showed that stereotypes could be made from such tin-foil matrices. Using steel punches he and Moore experimented with other materials, including lead, blotting paper and papier-mâché. Having exhausted his own means, Clephane obtained money from an unnamed gentleman in Baltimore for building a machine based on stereotypic principles. From the biography it is evident that Mergenthaler believed that stereotyping was not considered until the lithographic process had proved intractable, so Clephane probably confided in colleagues only on a 'need to know' basis. When he suggested building the stereotypic machine, Mergenthaler, who knew nothing of the process, doubted that it would work but, according to the biography, Clephane exclaimed, 'Give us the impression machine and let me attend to the rest!' The machine was intended to cast lines of type from a mould made by impressing one character at a time into a papier-mâché strip. It was completed towards the end of 1878, but once again the output was inconsistent and about a year later Mergenthaler left the project.

Mergenthaler and Hahl, who had agreed to payment by results, then asked for some tangible recognition of their services. This led to 'a little misunderstanding' which was resolved by giving each man three shares of stock, equal to one twenty-fourth of the interest in the National Machine Printing Company. [There is a slight inconsistency here; $28,000 is not exactly divisible by 24.] Mergenthaler had received no special payment for the patents that he had given to the company, but whether others would have thought his claim, or his attitude, as 'a little misunderstanding' is open to doubt. He was sarcastic about his share which he said was: 'a princely reward for his labours, considering that practically every patent on which the system depended was his invention' – and critical of Hahl's, 'there being not a single patent or invention to his credit.' Mergenthaler obviously placed little value on the resources that Hahl had provided for years, nor did he acknowledge his debt to Hahl for helping him to emigrate and avoid service in the German army. Work on the project continued night and day throughout the summer of 1879, but without success. When Mergenthaler said that the system seemed to be doomed, Clephane and his associates set up a small workshop in Washington which they finally abandoned in 1884. From 1879 until the start of 1883 Mergenthaler acted as an occasional consultant to the group.

Between 1876 and 1879 Mergenthaler was August Hahl's foreman and used his expertise to develop machines from Clephane's ideas. Without belittling Mergenthaler's contribution it must be recognised that these early devices were based on the concepts of another person rather than the outcome of original thought. However, there was great enthusiasm for the projects and much of the later experimental work being carried out on credit 'there was often not one cent of money left in the treasury of the [Hahl] Company to pay for the work.' After he left Clephane's project Mergenthaler became obsessed with the problem of inventing a successful typesetting system. He drew up a new plan which he:

> ... destroyed in a fit of anger brought about by the extreme financial straits into which he and the Hahl establishment had gotten themselves ...

All this early work on Clephane's projects should be considered as pure research. Despite intensive work and steadily mounting costs, none of the experimental machines had been developed into a successful product. It is also important to remember that Mergenthaler had always worked in a family group where success of the business depended on mutual trust and disagreements were settled better by compromise than by confrontation.

In the summer of 1881 Mergenthaler sold his entire interest in the National Machine Printing Company to a friend for $60 and some time later Hahl sold his holding for $900. Mergenthaler noted that by 1898 those interests would be worth hundreds of thousands.

In retrospect there were also good times in 1881. August Hahl made Ottmar his partner on the first of January and Ottmar married 20-year-old Emma Lachenmayer on 9 September. They had met at the Liederkrantz, a German singing society in Baltimore. Emma was also of German descent, the daughter of architect and sculptor Louis Karl Lachenmayer and Pauline Rosina Koener. They were married by a Methodist minister at the bride's home and there was such heavy rain that night that the guests were ferried by wagon from the street car terminal. During the early years of their marriage, when money was short, they lived over a confectioner's shop and were given a custard pie each week. Mergenthaler was very fond of sparrow pie and shot his own sparrows. Emma still played with dolls after their marriage and sometimes she had forgotten to make dinner before Ottmar came home in the evening but it was a harmonious union and their first four children, all boys, were born at this address.

## 5  *Mergenthaler as the driving force*

On 1 January 1883 Mergenthaler assumed the role of initiator for the first time. He dissolved his partnership, possibly to protect Hahl from investing in a risky venture, but more probably, to show that he was no longer a junior partner. He started a business in a small workshop at 12 Bank Lane, Baltimore, and approached Clephane with a design for a new stereotyping machine that would assemble patrices (punches) in complete lines. These would produce more even impressions in the papier-mâché stereo mould, or matrix, than those produced by machines that impressed one character at a time. The patrices were mounted on long tapering strips, or 'bands', each carrying a complete alphabet of upper and lower case letters, spaces, punctuation marks and numerals arranged in order of character width. This machine is now known as Mergenthaler's first band machine. [Latin scholars puzzled by the word patrix should note that it is defined in the Oxford Dictionary as a correlative term to matrix (female die). It cites *The Times* of 24 March 1883 as the origin of patrix. Therefore the word might have been coined by Mergenthaler or one of his associates.]

In 1882 Clephane had found a sponsor in Lemon G. Hine, a prominent Washington lawyer, who with Mr Frank Hume had bought the majority of the National Machine Printing Company stock. Mergenthaler thoroughly approved of Hine, describing him as 'liberal almost to a fault' and always ready to listen to the opinions of his experts. In January 1883, Hine gave the order to build the new machine and he, together with Frank Hume and Kurtz Johnson, paid for all the work done by Mergenthaler until it was consolidated with the other interests about a year later. This must have cost $14,000 according to an agreement subsequently drawn up between F. Hume & Others and The National Typographic Company. Hine was to give Mergenthaler 'some fair share in the invention' if the project were successful, the details to be decided entirely by Hine. As usual, a small experimental machine was made to demonstrate the potential advantages of the new system. The full size machine was tested towards the end of 1883. It produced good impressions but failed because the matrix took so long to dry that it made the process too slow. Mergenthaler decided that the matrix should be made of metal rather than papier-mâché, but that would require steel punches which would make it too expensive.

During this time, Hine put the project on a sound financial basis by setting up the National Typographic Company of West Virginia, which absorbed many smaller companies that had financed the earlier research. It was incorporated on 10 December 1883 with a share value of $1,000,000, consisting of 40,000 shares with a par value of $25 each. The original directors, all members of the so-called 'Washington Group' of investors, included Hine, President; Frank Hume; Kurtz Johnson; James O. Clephane and Abner Greenleaf. A list of enterprises that were incorporated into the National Typographic Company named some of the people involved; it included A. Hahl and Miss J. Julia Camp, a gifted typist who helped to test Mergenthaler's machines, but did not mention Ottmar Mergenthaler.

While trying to solve problems associated with the first band machine Mergenthaler went to Washington for a meeting with Hine and Clephane. He stated that the idea of casting lines directly from matrices stamped into the bands 'flashed across his mind' on the train journey, but Henry Lewis Bullen, the Australian printing historian, writing in the *Inland Printer* in February 1924, claimed that one of Mergenthaler's friends had told him that Hine had actually made the suggestion.

Regardless of who originally had the idea, it seems to have been the first proposal for a machine to combine composing and casting and it was built by Mergenthaler. In this machine, called the 'second' band machine because it also used bands, the operator pressed keys to set stops that caught each band at the correct character as it dropped from its initial position. After the line was cast the bands were restored to their starting point. Figure 6 is a diagram of this machine with the detail of a band on the left. Hine immediately ordered two machines to be built on this new principle; the first model being ready for trial in July 1884.

Stilson Hutchins, owner of the *Washington Post*, (see biographical notes) was among at least a dozen interested spectators who attended the trial of this machine at the Bank Lane shop. They arrived several hours too early and watched as Mergenthaler worked calmly until he was ready to start the demonstration.

In the biography the inventor claimed that he had composed and cast the first few lines, but another report suggested that his 17 year-old assistant Ferdinand J. Wich, set the first line although it was probably cast by Mergenthaler. Harry G. Leland wrote to Joseph T. Mackey, President of the Mergenthaler Linotype Company on 10 December 1938:

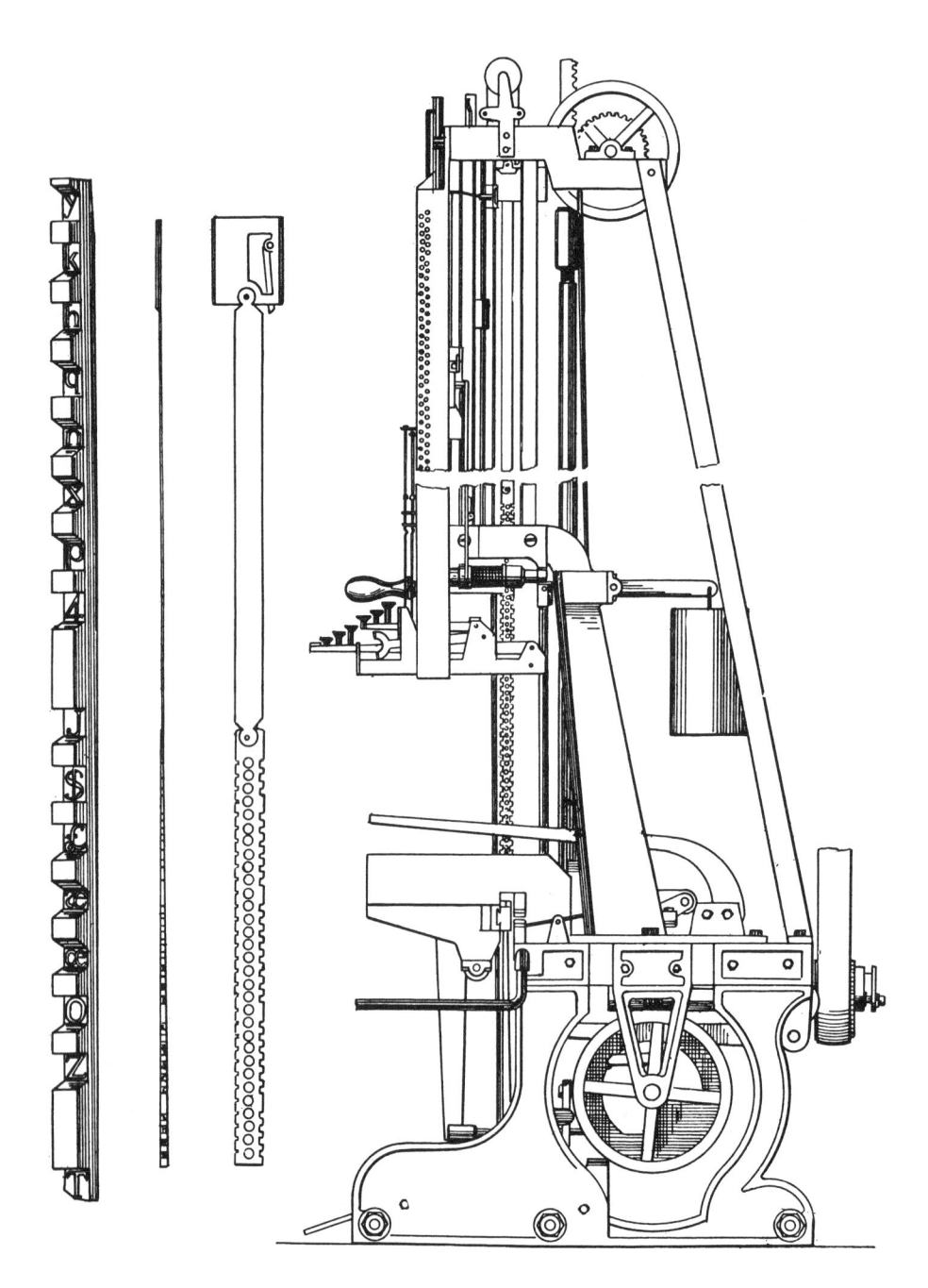

Figure 6  The second band machine including detail of band (on left)
*Source: Mergenthaler's Biography*

> ... Mr Mergenthaler told Wich to set a line, and he set up his sister's name and address. He then turned the machine over and the slug came out. After looking it over and studying it, Mr Mergenthaler threw it on the floor. Wich picked it up and retained it, thereby having as his own the very first slug from a Mergenthaler machine of any kind.

Enclosed with this letter was Wich's obituary from the *Baltimore News* of 5 December 1938, which recalled that Wich had installed one of the first Linotypes in England.

In the biography, Mergenthaler referred particularly to the operating skills of Miss J. Julia Camp who always obtained better results than other operators. The test was a great success, but the operator had to justify lines manually. Therefore it was decided to install automatic justification by wedges on the second machine. Mergenthaler found that his automatic justification mechanism would infringe Merritt Gally's 1872 patent for a single-wedge spacer so, in 1884, his backers bought the rights for $1,000 plus a royalty on each machine. Gally reported that a Mergenthaler representative told him a pitiful story of the poor German in Baltimore who had spent all his money and all he could raise from friends, but found he could not get a patent because of the priority of the Gally invention.

Although radically different in appearance and function from the machines used for newspaper production and not named as such, this was really the first Linotype. It was estimated that the machine would have to cost less than $400 to be a commercial success.

At this point, when Mergenthaler was on the threshold of success, it is important to understand what he had done. He had replaced the setting of movable type by hand with a machine that assembled matrices from which to cast complete lines of type. Each line consisted of individual character matrices, spaced and justified to a specified length. The concept of assembling matrices to make stereotypes directly was not original, but the inventor was probably unaware of the prior inventions. Over 80 years before the second band machine, Herhan obtained French patent No 285 for composing lines of matrices but his idea, which was not developed commercially, was to make a stereotype directly from a page of character matrices assembled by hand. The novelty of Mergenthaler's invention was in the mechanism by which an operator assembled and justified lines of matrices from which lines of type were cast and which were automatically restored for the matrices to be used to set the next line.

The successful demonstration of the second band machine changed the emphasis from research to development. Having accepted the principle of casting lines of type, The National Typographic Company decided to set up a machine shop at 201 Camden Street, Baltimore and Mergenthaler was asked to manage the works, but the tone in which the offer was reported in the biography and subsequent correspondence showed that he did not want to accept the post. He probably feared that the appointment would change his status within the enterprise from 'initiator and partner' to 'employee'. His agreement with the company, drawn up on 13 November 1884, signed by Mergenthaler and Hine and witnessed by Abner Greenleaf, promised him absolute management control of the works, a salary of $3,000 for one year, and 'ten per centum on the cost of manufacture of all machines from which the Company shall derive a revenue'. A third clause stated that if he left the Company after one year all his past and future inventions 'shall be the property of the Company, and be assigned to it by him, and in consideration therefor, the Company shall pay him ten per centum of the cost of all machines, devices or appliances manufactured by or for it of which any of his inventions or improvements or any part of them may form a part.' In return he promised to devote all his efforts and ingenuity to the interests of the Company. He was also promised one thousand $25 shares if he produced 'a machine fit for practical work'. In the first part of this agreement the percentage added to the salary was for machines from which the Company derived a revenue; the third clause mentioned the cost of manufacturing machines, but did not mention sales.

Meanwhile, Stilson Hutchins had become a prominent stockholder and promoter of the Linotype. In February 1885 he organised demonstrations of the full scale machine at the Chamberlain Hotel in Washington, DC, and gave a banquet at which the guests included Chester Arthur, then President of the United States, and other members of the government. During his short speech at the dinner, Mergenthaler said:

> I am convinced, gentlemen, that unless some method of printing can be designed which requires no type at all, the method embodied in our invention will be the one used in the future; not alone because it is cheaper, but mainly because it is destined to secure superior quality.

This claim was not realistic because the machine was still experimental. Hand-cut punches, which can never be copied exactly, were used to make the matrices in the bands. Each job would have been cast in new type, but the variations between different punches of the same character would have been unacceptable. Mergenthaler was also guilty of false modesty when he praised the gentlemen of the Board and minimised his own contribution.

The press does not seem to have reported the above events at the time although Goble assumed that news about a novel type composing machine probably spread widely among printers by word of mouth. However, there were some reports, even in the British trade press. In October 1884, the *Printers' Register* reported, without naming the source: 'An American contemporary says a new type-setting and manufacturing machine is attracting much attention in Washington.' It mentioned that $50,000 had been spent so far, that Mr Rounds, the Government printer, had declared (possibly referring to Miss J. Julia Camp) that 'a girl with sense enough to run a type-writer can set up as much as ten men now can, and make her own type as she goes along, too,' and that it cast lines of type 'at the stroke of a finger.' These details could apply to the demonstration at the Bank Lane workshop. On Saturday, 23 May 1885, the *New York Sun*, with the date line Washington, 14 May, identified Mergenthaler, The National Typographic Company and the banquet but, instead of February, gave the date as 'last fall'. This discrepancy might have been caused by a reporter writing 'Feb' badly and the compositor misreading it as 'Fall'. The report noted that the shares had risen and that a company of Northern capitalists who hoped to buy a controlling interest had broken off negotiations when the Printers' Composing Machine Company (limited) of Philadelphia, claimed that Mergenthaler had copied their machine. The *Printers' Register* carried a similar report in August 1885. However, there was no listing for such a company in the Philadelphia city directories for the years 1884 to 1886.

## 6 Portrait of an inventor

At this stage Mergenthaler comes across as a very ambitious skilled craftsman who adapted quickly to his new country and nationality. He apparently had inexhaustible good health and energy and his creativity had been stimulated by working with inventors at the Hahl shop. However, his plus points must be balanced against his negative traits. He was envious of the credit and material rewards given to others, in particular August Hahl, to whom he was indebted for being in the USA rather than in the German army in 1872. He was also hypocritical in being falsely modest about his name and achievements when he spoke at the dinner at the Chamberlain Hotel, while in fact he was always pushing for recognition. Occasionally he did acknowledge the efforts of others: he gave credit to Clephane for his originality, which was evident by Clephane's many patents; to Hine for providing cash and delegating responsibility to his experts, although he did not mention that Hine may have suggested the concept of direct casting; finally he gave credit to Miss J. Julia Camp as the most competent operator.

Mergenthaler was a steady precise workman who tended to be cautious rather than rash but was sometimes reckless and intemperate. He claimed that having no printing background had been an asset because he was not bound by preconceived notions. He said that he had surveyed the field to avoid making the mistakes of others but his survey was probably limited to inventions registered at the United States Patent Office since he was obviously unaware of Herhan's French patent. He was overconfident and reckless to invite people to a demonstration of an untested machine and overoptimistic in setting the date before he knew that the machine would be ready but, having got an audience, he did not let it faze him. Why did he only give a brief glance at the first slug and throw it on the floor? That was unnecessarily casual! He showed an impetuous and irrational streak under stress by selling his shares for a pittance and destroying plans in a fit of temper. These actions may have led to later rumours of madness which were widely held, even in living memory. His experience with the neighbour who belittled his invention of the lathe chuck by implication showed him to be very sensitive and easily hurt. The memory obviously rankled throughout his life.

The terms of the agreement between Mergenthaler and The National Typographic Company of West Virginia were supposed to have been left entirely to Hine, who was described as a prominent Washington Lawyer. However, that document seems to have been badly drafted. There were inconsistencies in the conditions for paying the ten per cent royalty, between the first part of the agreement and clause 3. Contention could arise over defining 'cost of manufacture' which could consist of materials and labour, but might also include factors like buildings and plant. The wording of clause 3 that: 'the Company shall pay him ten per centum of the cost of all machines, devices or appliances manufactured by or for it of which any of his inventions or improvements or any part of them may form a part' was always open to dispute, particularly if the machine were enhanced by others. In the biography Mergenthaler claimed that the ten per cent royalty was 'small indeed' and depended 'entirely on his success in producing a commercially profitable machine'. He maintained, naming himself as 'the inventor', that he welcomed these terms because he could not be accused of trying to make quick profits when he continually tried to perfect the product. He did not consider the lack of returns on investment for the stockholders to be as important as his loss of expected royalties and conveniently overlooked the fact that he was paid an attractive salary and bore none of the development costs that were mounting steadily while he continued to experiment.

# ❦ TWO ❦

# *The beginning of hot metal composition in newspapers*

The mid 1880s were an exciting time for the United States of America. The Civil War had ended some twenty years earlier and the Centennial of 1876 showed that the country was becoming an industrial power. New York City was starting to develop its famous skyline. In 1875, Whitelaw Reid (see biographical notes) owner and editor-in-chief of the *New York Tribune*, had commissioned the new *Tribune* building, the first skyscraper in that city. Just down the road from the paper the Brooklyn Bridge was opened in 1883 and three years later the Statue of Liberty was erected on Bedloe's Island at the entrance to New York harbour.

## 1  *Refining and redesigning*

The successful demonstration of Mergenthaler's second band machine at the Chamberlain Hotel established the principle of the single operator line casting machine. Once this technology had been accepted all further work was aimed at improving the mechanism for presenting justified lines of matrices to the caster and improving the quality of the slugs.

However, Mergenthaler recognised that there were serious drawbacks in the design. The matrix bands were not true enough; the machine could not do tabular work; the operator could not see what was being set and could not correct errors made during assembly; and it was not possible to set characters that were not on the bands. To be a commercial success the machine had to produce a printing surface that was acceptable to the trade. He realised that a machine based on the circulation of individual matrices would overcome the above problems. The operator would be able to see the matrices being assembled and would be able to do limited tabulation; matrices of each character would be liable to similar amounts of wear as they circulated in the machine; and it would be possible to correct errors and, if necessary, insert special characters by hand.

The investors were upset when told about these problems. According to the biography: 'This was a surprise to them, and a big one too, but not a pleasant one. The new idea called for the construction of an entirely new machine, and new machines were commencing to become odious to the promoters.' Hine said, 'Not many stockholders can stand being told that we have the best machine in the world, but we are going to make another which is still better.' They were looking for a return on investment and expected the second band machine, which could produce up to four lines a minute, to be the definitive version. Although the change of plans caused more delay, Mergenthaler convinced the investors that the new machine would be far superior to the previous model and should be developed.

The account of this invention in the biography suggests that the inventor got the idea of the single matrix machine quite suddenly, but this was not so. On 21 October 1884, over three months before the demonstration of the second band machine at the Chamberlain hotel, he had applied for a patent on a form of circulating matrix machine [US Patent No 317,828, granted 12 May 1885]. There seems to have been a complete blackout of information while the new machine was being built.

## 2   *The syndicate moves in*

The Hon Stilson Hutchins, a handsome, swash-buckling, devious man with an honest open countenance, was probably the first person to realise the potential value of the Linotype in newspaper production and but for his greed and drive the machine may never have got beyond the experimental stage. After the demonstrations at the Chamberlain Hotel he started to promote the machine in earnest and approached wealthy newspaper men with the aim of forming a syndicate to take over the project. His method is exemplified by correspondence in the Whitelaw Reid papers in the Library of Congress. Hutchins wanted Reid to join the syndicate because he was a prominent newspaperman with substantial resources and the *New York Tribune* had been non-union since 1877 following a pay dispute. He was an ideal prospect to exploit a machine with the potential to make four compositors out of five redundant.

On 10 February 1885, the month of the banquet, Hutchins sent Reid a two-page letter typed in capitals under the *Washington Post* masthead. Although the letter contained many contradictions, Reid must have taken it seriously. Hutchins opened with an apology that his employee, Eaton, had written to Reid without authority. Hutchins, who would not try to force Reid into anything, was by far the largest stockholder in the Company but had no stock to dispose of at any price liable to be offered. Having denied trying to sell stock, he opened his sales pitch by stating: 'We are anxious to enlist the interest and co-operation of such persons as yourself, and to induce you, and others like you to join, would sell you stock at lower than market rates, if indeed it can be said to have a market rate, nor should we expect you to bear the brunt of any fight necessary to introduce it. The machine can just as easily be put into job offices first, and there be made to show what it can do, as in a newspaper office. In such a case it would not impinge upon the union.'

Hutchins continued that he would either go up into a realm of opulence hitherto undreamed of, or down with this invention; he claimed to have a device that would revolutionise composing rooms in the printing business; further no other invention could possibly compete for the next ten years because their granted claims almost covered the entire field.

He enclosed examples of setting to show the sort of work being done, with the comment that the quality would improve when they were able to recruit men from the type founders to cut their punches. He expected costs to be about 5 cents per thousand ems, exclusive of royalty. The machine had already proved its reliability, and he was sure that it would be better than hand composition for setting late news in a daily paper. He claimed that his principal reason for writing was to assure Reid that nobody had tried to sell him stock for its profit, *with our consent*. He invited Reid to look at the machine carefully and critically with his associates, as it was not intended to put it out for commercial use until it was right.

He expected a number of machines to be at work in job offices within the next few months. He would put one in his own office were it not that he might face pressure from union printers at the Government Printing Office in Washington, DC. He did not know that they would fight it but did not want to give them the chance. Hutchins ended with the assurance: 'that there is more in it than all other enterprises now accessible to moderate capital' and an apology for the letter being typewritten: 'I was anxious to write you immediately, and could not steal the time to do it personally.'

In the biography, it was stated that the syndicate was led by Whitelaw Reid owner of the *New York Tribune*, and included W. N. Haldeman of the *Louisville Courier-Journal*, Victor Lawson and Melville Stone of the *Chicago News*, Henry Smith of the *Chicago Inter-Ocean*, W. H. Rand of Rand, McNally & Co, Chicago, and last but not least Stilson Hutchins, owner and founder of the *Washington Post*. There is an anomaly here; legal documents identify the *Chicago Inter-Ocean* as an early user of the Linotype, but correspondence from Reid showed that Henry Smith was not a member of the syndicate – it should have been William Henry Smith of the *Associated Press*, who became secretary of the National Typographic Company when the syndicate took over its management.

These men would have been familiar with machines that set movable type. Before investing in the enterprise, they and their agents paid many visits to the Camden Street works. Reid, who had Burr typesetters in his composing room, showed the greatest interest in the new machine. He spoke warmly to Mergenthaler and forecast that he would return to his native Germany a wealthy man with a castle on the Rhine, unaware that the inventor regarded himself as an American who did not intend to live in Germany. However, that phrase 'castle on the Rhine' was to torment Mergenthaler for the rest of his life.

There was considerable confusion about how the syndicate contrived to take over the management of the company. Mergenthaler thought that they had paid the record price of over $300,000 to buy a controlling interest – probably the highest ever paid in the USA for an invention that had not been proved in practice. This is untenable because the company's shares were over par. The syndicate could not have acquired stock worth over $500,000 for only $300,000. On the other hand, Clephane claimed that the syndicate could secure only 7,000 of the 40,000 shares at prices between eight and ten dollars over par and that, although the syndicate held only 17.5% of the stock, several prominent stockholders agreed to cooperate with them in the management of the company. Unknown to either man Melville Stone, on behalf of the syndicate, and Stilson Hutchins, signed an agreement on 14 March 1885 under which the syndicate agreed to buy 5,000 shares of National Typographic Company stock at $50 per share and Hutchins and his associates would hold and represent 15,100 shares; the whole 20,100, just over half the shares, to be used to control the organisation in perpetuity; subject to satisfactory practical tests, which were not to be publicised, and legal examination, both to be completed in 20 days from the date of agreement, and the shares to be paid for within 10 days thereafter. Subsequent events show that this fell through because no agreement had been reached within the time limit. On 17 February 1886, Reid wrote to Stone that they could finally close the business about the syndicate. The total number of shares of the National Typographic Co purchased was 7,000 at $32 per share, with associated expenses of $14,024. Another 1,050 shares that a member of the syndicate had bought at $33⅓, without any additional expenses, were added to the syndicate's purchase. This brought the total number of shares to 8,050 for $273,024, nearly $300,000 when rounded up, but there were several arithmetical inconsistencies in the letter. Thus, both Clephane and Mergenthaler knew something about the take over but were obviously unaware of the details. Having gained control for the syndicate, of which he was also a member, Hutchins was paid a $200,000 'finder's fee' as his reward for discovering the Mergenthaler machine.

The delay in the take over could have been due to accusations, such as that noted in Chapter 1, that Mergenthaler had copied another machine. Despite there being no reference to the Printers' Composing Machine Company (limited) of Philadelphia in the city directories, the syndicate consulted experts before deciding to support the project. It was claimed that Mergenthaler's automatic justifier infringed that of Jacobs W. Schuckers (see biographical notes) of Philadelphia, Pennsylvania who was described on his patent specification as Assignor to The Electric Typographic Company of New York, NY. The two justifiers were developed simultaneously, and probably independently, but Schuckers had filed his application on 27 February 1885, 49 days before Mergenthaler. Correspondence on this issue continued into the autumn.

Smith, who had consulted a Mr Dickerson about the Schuckers case, sent Reid a telegram on 28 May 1885, giving Dickerson's opinion that Schuckers's lawyer had been stupid! If the patent were based on the papers submitted by Schuckers there would be no infringement, but there could be interference if they were amended. On 6 June 1885, Smith noted that Stone had decided that Schuckers was of no account.

Hine wrote to Reid, on 18 September, that he had been offered a *majority* of the stock in the Schuckers patents, *in a very private way*, for $50,000, and wanted to know if there was any use in opening negotiations. Reid replied on 22 September, in an almost illegible letter, that he thought not – judging from Dickerson's opinion.

On 9 November, Reid wrote to Hine, that the Schuckers crowd claimed that the Mergenthaler machine infringed a justifying device which Schuckers had filed for patent last January and that they would be able to enjoin the National Typographic Company as soon as it began to use the machines. They claimed that this device was not on the machine when Mr Dickerson saw it. It was important to find out if this were true. Doubtless the patent attorney, Philip Tell Dodge (see biographical notes) could ascertain the facts.

Hine replied next day. Dodge said that it was impossible to learn with any certainty what application Schuckers had in the Patent Office. It was very improbable that he had filed an application last January, or even last April, that interfered with Mergenthaler's justifier, because an interference would almost certainly have been declared by now. Dodge did not seem impressed with Schuckers's claim. He would try to avert any danger from that source, especially to learn what Schuckers was trying to do.

At this point interest in Schuckers was apparently dropped. The experts thought he presented no immediate threat, possibly because his people had claimed that the application was filed in January, when the actual date was at the end of February, presumably *after* Schuckers might have seen the second band machine at the Chamberlain Hotel. However, Schuckers's application was amended. It was divided and filed again on 18 November 1886, the very thing that had worried Dickerson.

## 3  In-fighting in the syndicate

The new management team took over in the spring of 1885, with Stone as chairman of the executive committee. The biography made it clear that Mergenthaler resented Stone as inexperienced and lacking Hine's ability to manage the enterprise. The inventor was factory manager when he first met the syndicate and they treated him as a hired hand, rather than a major member of the venture. This was the attitude of Stone who tried to order him to move the factory to Chicago. Mergenthaler was reluctant to leave a well established factory and his circle of friends for the uncertainties of Chicago and wrote that he had politely but firmly declined. However, he was no diplomat and his *polite* refusal could well have been misconstrued, particularly by a dictatorial newspaper mogul who may not have thought that any refusal could be polite.

Reid and Smith were obviously close friends who broke with tradition by writing to each other on first name terms, but to others they observed normal nineteenth century formality. Their letters showed that they wanted to control the operation from New York. On 6 July 1885 Smith wrote that he would:

'rather the Mergenthaler outfit went to New York. I think we can manage it better there than with the help of others elsewhere.' In this instance they backed Mergenthaler – to thwart a move to Chicago.

After stepping down as chairman Hine became Reid's contact with the Baltimore factory. On 10 September, Reid sent him a draft resolution for the Board of Directors proposing that the main office of the company be established in New York City and mentioning the need to provide ways and means for carrying on the shops in Baltimore.

Stone must have resented Reid and Smith's support for Mergenthaler and, possibly out of pique, did not take his duties as chairman seriously. As a result, Reid became virtual leader of the Board. Stone had postponed meetings at short notice; had left Reid to convene meetings and then had to be cajoled into attending; and sometimes submitted reports too late to allow enough time for lawyers to scrutinise them before a meeting.

On 6 October, Reid wrote to Smith, generally approving Stone's draft report about introducing the machine. He thought that it should stress that it would not succeed unless they could interest people who had the facilities and were willing to take risks. Several excellent machines that did work about half as well as this promised, had failed to secure general support even without active hostility from the unions. He thought it vital for Smith and Stone to be in New York before the Washington meeting.

Although there was no publicity during this development period, Reid was in regular contact with other New York newspaper men, particularly Charles Dana of the *Sun* and Joseph Pulitzer of the *World*. Dana told Smith that he had had a long conversation with Mark Twain, on the subject of a Hartford typesetting machine in which Twain had a heavy interest. Twain assured him that the Washington machine was a failure and that in any case he was going to be in the market with his before the Washington people could possibly get ready. He said that his machine would set at the rate of 6,000 ems per hour, and that he would sweep the field. Smith said: 'Of course we don't believe this, but we fear that he may be able to make us a good deal of trouble if we have much delay.' Reid's reaction to this bluster was to tell Hine confidentially that he was averse to showing off the machine widely, because 'Clemens [Twain] is a remorseless man in money matters, and, his enemies say, not over scrupulous.'

Early in 1886 when Reid considered the possible problems associated with introducing the machine, Pulitzer warned him of 'riots in plenty', but in the event there was no violence.

## 4  Breakthrough – the prototype of the single matrix machine

In the summer of 1885, a prototype of the single (or independent) matrix machine was tested and overcame the drawbacks of the previous models. The operator could correct a line during assembly, set tabular matter, and insert italics, or other extraneous sorts, by hand. This machine, later nicknamed the 'Blower', required a blast of compressed air to shoot the matrices along a horizontal path into the assembler. Figure 7, is an artist's impression of the prototype.

There were no public progress reports at that time, but letters were regularly exchanged between the members of the Board. On 10 June, Smith telegraphed Reid: 'Machine will be furnished July first. Stone has examined it and thinks it remarkable.' This was less than six months after the demonstrations at the Chamberlain Hotel; so soon that Mergenthaler must have been designing and building the machine at the same time. It was tested and improved during the next few months.

On 10 September, Reid asked Hine for the latest news of the Baltimore machine and Hine replied in a long letter dated 18 September, that it was in a very encouraging condition. He expected it to be ready for demonstration early the following month. Smith sent Reid a pencilled note, on 26 September, telling him that he was impressed with the latest version.

On 29 September, Reid sent Smith a confidential letter saying that he would try to test the machine as soon as there was a chance and did Smith think it wise to print a page from it, before final arrangements were completed. Reid thought that it might be playing into their opponents' hands; presumably he was concerned about the Schuckers group. Three days later Reid wrote to Smith to tell him that the machine would be ready the following Wednesday [7 October 1885] and following this the syndicate decided to start large scale manufacture.

## 5  The Mergenthaler Printing Company

The National Typographic Company had no funds to finance this effort, so it was decided to organise The Mergenthaler Printing Company, a new corporation with a stock value of $1,000,000. Whitelaw Reid was appointed President and General Manager. There is no record of how or why it was named after Mergenthaler who was never a company officer. Stock in the new company was not offered to the public. Stockholders in the National Typographic Company were offered as many shares in the new company as they held in the old, any allocations not taken up were to be divided among existing stockholders willing to invest. This seems to have been a means for the syndicate to obtain a majority holding in the new company. Hine realised the potential danger and encouraged the Washington group of stockholders to take up their options of which: 'It is not thought that more than twenty-five per cent of the capital will ever be called for.' Always a proud man, Mergenthaler felt honour bound to take up his option of 1,000 shares but could not raise the $25,000 assessment because all his assets were tied up in his business. He asked the directors to accept his subscription and let him pay for it from his royalties, but this was refused on legal grounds. In despair he made the disastrous decision to accept a loan from Reid to buy his shares. The terms were quite harsh: Reid held the shares as security together with an irrevocable proxy and charged 6% interest. If the total $25,000 were called for, the interest of $1,500 on this loan would have been half

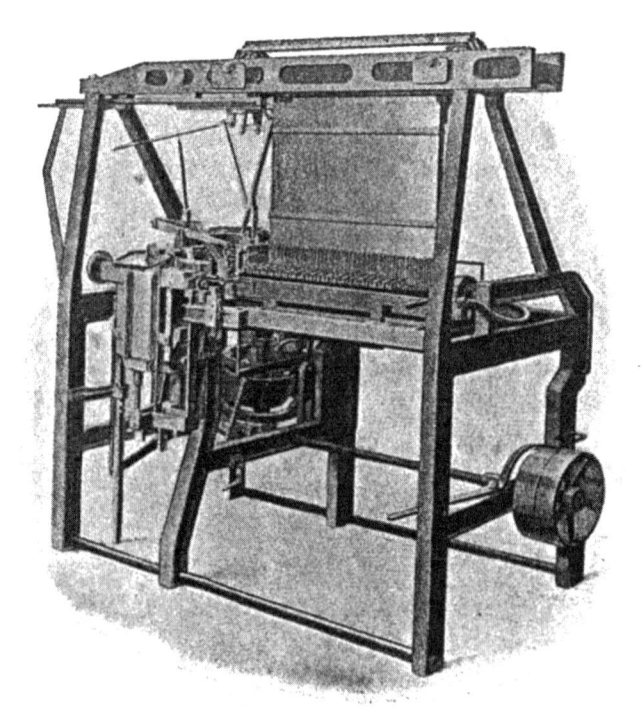

Figure 7   An artist's impression of the prototype of the Blower Linotype
*Source: The L&M News, May 1928, p 11*

his gross salary as manager, before he could start to pay off the principal. Mergenthaler did not seem to realise the magnitude of this commitment and warmly thanked Reid for subscribing for his stock. On 12 January 1886, Mergenthaler sent the Contract and Proxy to Reid with the comment: '[the] general arrangement of the matter is very satisfactory to me, and I feel satisfied that when the obligation is due our enterprise will be in a condition which will enable me to meet the same.' When the company called for a subscription of 20% Mergenthaler was immediately in debt for $5,000.

In the biography Mergenthaler claimed that funds were now plentiful and that he was told to get into production as quickly as possible 'without undue regard for economy.' He engaged Mr Sumter Black to revise the working drawings, one of the few times that he actually delegated a task, but was appalled when the draughtsman told him that it was a six-month job. When Mergenthaler said that he had built the machine and made his own working drawings in less than six months, Black replied that those drawings were adequate for an inventor producing one machine but lacked the detail necessary for others to produce them in quantity. Later Mergenthaler acknowledged that Black was right.

Reid reminded Hutchins, on 5 February 1886, that he had not paid his subscription, which was to be one-eighth of the syndicate, saying in effect: 'Pay up or leave the syndicate!' He said: 'We are as anxious as anybody to make money out of this great invention, but we want to make it out of the machine, rather than out of stock speculation.'

Meanwhile Reid who had sufficient resources for long term investment had bought all the available stock in the Mergenthaler Printing Company at par, but he cautioned Mr William Walter Phelps on 26 February that investment in the company was purely speculative. The rest of the capital would probably be called up that year and, instead of trying to sell it at a premium, people would be hunting around for somebody to save it from being confiscated under the law for non-payment of calls. There was not a ghost of a chance for any return that year nor could anybody tell when there would be. The machines now under construction would not be finished on time and they did not dare to order others, because there were many untried devices which had to be tested before they could launch out on any large scale.

When they were sure that the machines would work they would make a number for members of the syndicate. Then a factory would be bought or rented and equipped for large scale manufacture. He believed the machine would be profitable, but it was still very speculative and outsiders who fooled with it were likely as not to regret the same.

This was the advice he would give to anyone in Phelps's position if approached on the subject. Meantime, if he could get any of the new stock for the assessment already paid and interest on the same, he would do it but would not pay a cent beyond that. It would be to the syndicate's advantage to have a couple of thousand more shares to make absolutely sure of control, otherwise he would not propose buying any at present.

## 6  The first Linotype is produced

On 30 November 1885 Reid wrote to Stone that the working drawings would be completed within two weeks and that as soon as they were ready they should start to organise contracts for manufacturing the first hundred machines. Mergenthaler earnestly objected to the enormous risk involved in such a large order and the board agreed to reduce the number to twelve to be delivered by 1 April 1886.

Reid sent Thomas Miller, his best machine operator, to Baltimore, to work on the machine for a week, but he was constantly interrupted by the draughtsman and by Mergenthaler making little changes as he observed a printer working the machine. Miller estimated that he could set 5,000 ems per hour on it, where a top hand compositor could not normally maintain more than 1,000 ems per hour. However, Reid's machinist, Mr Thompson, believed that he could get up to 6,000 ems. Reid intended to send Miller back for another week of what he hoped would prove steadier work after Mergenthaler had finished 'two or three days tinkering' and wrote to him on Saturday 5 December 1885: 'Will you be ready for Miller on Monday next with a fair chance for continuous work?' It was followed by another letter dated 7 December that Miller would report for work Tuesday morning. Yet another letter of the same date said that Miller would report on Wednesday morning instead of Tuesday; to which the inventor replied by Western Union Telegram: 'Machine working all right; will expect Miller tomorrow morning.'

The experimental output was printed in the *New York Tribune*, the only perceptible difference in appearance being that the lines of the type were finer and sharper. There was no comment in the paper. On 14 December Hutchins asked Reid for a copy of each issue of the *Tribune* that contained an article printed from National Typographical slugs. Next day Reid replied that they would try to make a file of marked copies of the papers containing articles set on the Baltimore machine to send to him.

By this time the Remington typewriter had been on the market for over ten years and many typists were familiar with the QWERTY keyboard. Reid wrote to Mergenthaler on 29 December 1885, that Clephane had said that recent improvements would allow him to use the typewriter keyboard on the machine. If using the typewriter keyboard involved no mechanical difficulties and would not delay matters it would be a great advantage over the current keyboard. It would give them the benefit at the outset of a large body of operators who could be called upon in case of need. Mergenthaler replied by return of post that Clephane was mistaken. The improvements to the keyboard related to speed and reliability, and it was not convenient to apply the typewriter keyboard to the machine.

Mergenthaler was under great pressure to manufacture the machines on time and, on 8 January 1886, asked Reid to excuse him from the election meeting the following week, because they were making improvements to the machine that required his personal supervision. In general he was confident about progress on the machine, but was having problems with the matrices. These were flat pieces of metal $1\frac{1}{4}$ inches [32 mm] long by $\frac{13}{16}$ inches [21 mm] wide with the mould of a character on the edge.

Those matrices were vital. Reid obviously wanted the output from the machines at the *Tribune* to match the movable type used on the paper so that he could use them as much, or as little, as he chose without having to make the matter public. [In Part III of the 1989 edition of the biography Carl Schlesinger described the difficulty of identifying those parts of the paper set by Linotype from those set in movable type. He decided that the matrices must have been electrotyped from fonts of movable type.] To protect the newspaper from adverse comment, in case the Linotype failed, Reid had the backup of three Burr typesetting machines for setting movable type and the case hands in the composing room to produce the paper as usual.

Time was very tight because by 23 December 1885 Mergenthaler had not received the type to use as a pattern. By early February 1886 matters were becoming desperate. Mergenthaler started to investigate a process for casting matrices, because electrotyped matrices wore out quickly, but had to find an alloy that would repel type metal and would not deform under heat or pressure.

Reid wrote to his associates and Mergenthaler that he was getting a little nervous about the delay on account of the matrices. He suggested that electrotyping in nickel should continue as before in order to have some matrices ready for the first of March whether or not the new experiments were successful. In rather too diplomatic terms he told Mergenthaler to get the basic design working before looking for improvements. He was not averse to experiments – but not at the cost of any delay to the machines.

Towards the end of February 1886 Mergenthaler said that he had found just the metal for casting matrices and on 9 April 1886, more than a week after the deadline for machines, wrote to Reid: 'Work on the new machines is being carried on vigorously. The first one is just now being built up and commences to look like what it is going to be. Everything so far goes together nicely. Hoping to be soon able to inform you of the completion of a perfect working machine with *cast* matrices.'

By mid-April Reid's growing frustration showed in his correspondence to others. In a letter to Stilson Hutchins he wrote: 'You are certainly right as to the necessity of getting one of those machines either into our office or some other office at the earliest practicable moment. We were promised, within a fortnight, that one of the machines should be ready to have steam turned on this week. The last letter from Mr Mergenthaler spoke of a little delay, but evidently he was not expecting it to be long. Just as soon as this machine is running, I hope to be able to make [some] sort of move in the direction you suggest.'

However, no machine was delivered and Reid wrote to Mergenthaler on 6 May: 'You don't need to be told that it is a very great disappointment, both to the Washington stockholders and what is called the New York syndicate, that the first of May finds us a month after the date fixed and still without a single one of the twelve machines ready for work. From the report sent to Mr Smith, I gather that one of these machines might now be ready for work if the course strongly argued last February had been taken and the preparation of matrices had then been begun by the old electrotyping process. Quite probably these matrices could not have lasted more than two or three months, but meantime the machine would have been at work and you would have had further leisure for perfecting your new plan for casting matrices.'

Reid kept in constant touch with Baltimore and, on 17 May, wrote to Hine: 'According to reports from Baltimore, one of the new machines is now in good running order and needs only an adequate supply of matrices to be set at once at regular work.' Hine replied next day: 'that a complete font of type matrices will be ready and the machine running tomorrow. There will probably be some filing and grinding needed before all the parts will work smoothly together but this will not take more than two or three days. I think we had better keep the machine running in the shop for about a week before removing it. I can see no possibility of any failure or weakness in any of its parts but it seems a little cruel and dangerous to remove it from maternal care the first week of its existence.'

Hine continued, no doubt hoping to placate Reid, that: 'the men work very industriously and take a real interest in the work. The trouble is – if it ought to be called a trouble – that Mr Mergenthaler is a characteristic inventive genius with astonishing fertility of resources in mechanics and a mind too original to adopt suggestions in his work. In short he is a pleasant estimable German. I have not found the loss of anything by accident. Mergenthaler is the last to leave and carefully goes over the factory every night. There is nothing that the thief could make pay for the labour and risk of stealing and organised distructionists have not yet considered us.'

On Tuesday 22 June, Smith wrote to Reid about delivering the first machine. Mergenthaler had kept the men at work until eleven o'clock the previous night and the machine could now be shipped on Friday with a full set of matrices. He had given strict orders to arrange with the Pennsylvania Central Rail Road Company for the prompt delivery of the machine in New York on Saturday morning [26 June]. Mergenthaler would go up on Friday night. It would not take long to set up the machine and connect with the gearing which, it was assumed, would be ready in advance.

Reid evidently had planned to leave New York on Thursday 1 July, and Smith tried to reassure him with the statement: 'I could have had the machine packed and shipped Wednesday, but nothing would have been gained so far as your getting off on the 1st is concerned, and something would have been lost with Mergenthaler. I found him really ill, and I have been worrying all the time lest he break down altogether. He has trouble with his lungs. I have therefore taken that course that will avoid irritating him, and yet promises to get you off on the day you have fixed.' This seems to have been the first reference to Mergenthaler's failing health.

Reid replied the next day that he would try to have shafting ready by Friday night, but would not attempt to arrange blower and air-blast until Mergenthaler arrived. He also wanted to know the size and weight of the machine and suggested that it could be hoisted like a safe if it would go through a window four feet eight inches wide. If it had to be taken apart any pieces weighing less than three thousand pounds and measuring less than two feet six inches could be brought up by the rear elevator.

Reid made arrangements with a safe-hoisting company to collect the machine from the Jersey City freight depot and install it in the *Tribune* building, but the shipping date had slipped because on Monday 28 June, he sent the following message to the Freight Office: 'Heavy machinery direct to me at Tribune office, New York, was shipped from Baltimore yesterday afternoon. Bearer has contracted to take charge of it and hoist to its place in my office. Please deliver to him, holding this as your voucher, and oblige.'

The delay also affected Reid's plans to leave New York on 1 July. Part of a letter of that date stated that Reid would be in the office on Friday and Saturday [July 2 and 3] between one and five o'clock. There was no further correspondence about the installation date of the first machine.

## 7   The first machine goes to work

In the biography, Mergenthaler glossed over the first installation in a sentence: 'In July, 1886 we find the first of these machines completed and at once forwarded to the composing room of the *New York Tribune*, where it was used on the daily paper and also to set a large book called *The Tribune Book of Open Air Sports* which book was composed entirely by this first linotype machine that ever went into commercial use.' As mentioned above, he seemed oblivious to the urgency and drama of that first installation.

In the August 1936 issue of *Linotype News*, John T. Miller, the first Linotype operator on the *Tribune*, aged only 20 at the time, recalled the arrival of the first machine on 3 July 1886: 'Through the Frankfort Street end of the composing room ran a huge flagpole. To that pole, that memorable afternoon in 1886, a block and tackle was attached. Nine stories below, in the street, was the first Linotype. It was completely boxed, and on top of it stood a man to guide it through the network of telegraph wires. [A safe rigger was paid $35 for this service!] The middle one of our three Burr machines had been shunted aside to make way for the new device, and the Linotype was slowly drawn up to the composing room. As the man who made the dangerous ascent threw one leg over the large window sill, he gave loud expression of his relief, and soon Ottmar Mergenthaler and Charlie Letsch had the Linotype in working order.'

'Naturally, the arrival of the new and improved contraption made a big stir. All of us who could manage to, crowded about for a good look at it. Most of the hand men didn't think much of it, and many said so. Even we Burr operators were a little skeptical at first. But we were the ones, we knew, who'd have to handle it, and I was determined to learn all I could about it as quickly as possible.'

Despite reports of a crowd when the first Linotype was delivered there was no publicity about the installation of the machine. The first known article and illustration released to the general public appeared in the *Scientific American* on 9 March 1889, nearly three years after the first Linotype came to the *Tribune*. The front page illustration, much reduced, is shown in figure 8.

The machine went into production the same day, but it was too late to set matter for that day's edition. John Miller explained how Linotype slugs came to be in the *Tribune* on 3 July 1886: 'Those slugs were cast by my brother Thomas Miller, now dead, who worked for some time with Mergenthaler in Baltimore, I clearly remember when the slugs arrived, and I distinctly recall that the postage on the package amounted to ninety-five cents.' Every day, after the paper had been put to bed, the Linotype was modified to compose the 500-page *Tribune Book of Open-Air Sports*. This was a major task; it took an hour to change from news to book setting.

Data based on personal recall 50 years after the event is always open to doubt. In this case, popular belief that the machine was delivered and installed on 3 July, is disproved by Whitelaw Reid's correspondence books in the Library of Congress. A letter to his wife, dated **2 July 1886**, stated: 'Machine here and working beautifully. On farm next Thursday.' A memo dated 3 July 1886, confirmed that he would leave New York next Wednesday [7 July] – but made no mention of the new machine.

When asked about the new machine Reid refused to give details until it had been thoroughly tested. On 9 September 1886, his secretary wrote to Henry P. Hay, First Auditor's Office, Washington, to say that the new 'type-setting machine' is doubtless the machine for casting 'linotypes' from moulds of letters. This machine, the property of the National Typographic Co and the invention of Ottmar Mergenthaler, is not yet on the market, and is still being perfected. When ready for introduction, you will have no difficulty in getting particulars. This was possibly the first use of the word linotype, but applied to the printing surface, not the machine. Reid sent a similar reply to a John Doran on 17 September 1886.

In November 1886, the *Tribune Book of Open Air Sports* was offered for sale to *Tribune* readers, by subscription only. A review in December, after the lists had closed, stated that the print was clear and attractive and that it was the first product in book form of the Mergenthaler machine which wholly supersedes

# SCIENTIFIC AMERICAN

A WEEKLY JOURNAL OF PRACTICAL INFORMATION, ART, SCIENCE, MECHANICS, CHEMISTRY, AND MANUFACTURES.

Vol. LX.—No. 10.
Established 1845.

NEW YORK, MARCH 9, 1889.

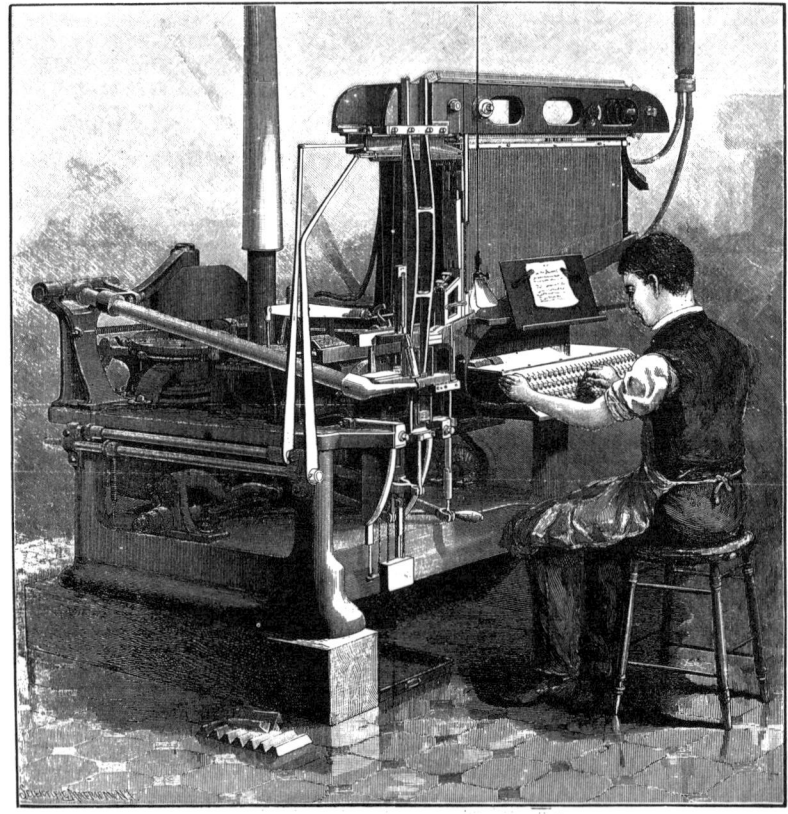

$3.00 A YEAR.
Weekly.

**A MACHINE TO SUPERSEDE TYPESETTING.**

Prior to January 1, there had been issued from the U. S. Patent Office upward of 160 patents relating to typesetting and type-distributing machines. All such devices, with many others known only in foreign countries, have thus far, however, met with but little favor among printers, and they have not been employed in practical work to a sufficient extent to have any appreciable effect in this most important branch of the printing business. Printing presses have been improved almost beyond comparison with those of the earlier days of the craft—when only about 200 impressions were obtainable per hour from small forms, as against more than 20,000 copies now made per hour of our largest newspapers; but the typesetting part of the making of books and newspapers has remained substantially where it was left by the earliest users of movable types.

The accompanying illustration represents the latest, and in many respects the most remarkable, of the numerous machines which inventors and mechanics have from time to time devised in their long-continued efforts to find some practical means by which to supersede or cut short the tedious work of typesetting. It is known as the Linotype machine, from the nature of its product, but would probably be more generally designated as the "Tribune" machine, from the fact that it has been in practical use in the New York *Tribune* office for more than two years, where it now does substantially all the work formerly done by the compositors of that paper.

It is not, strictly speaking, a typesetting machine, but forms type bars, each of the length, width, and height of a line of type, and the exact counterpart of that which a compositor would set up, except that each line is formed of one entire piece of metal, instead of as many different pieces as there are characters, spaces, etc. A representation of such type bar or slug is given in one of the small views. The key-board in front of which the operator sits has 107 keys, each marked for a capital or lower case character of a fount of type, or for the figures, points, or compound letters used in connection therewith, many of the letters most frequently used having several keys. The operative parts are carried by a rigid metal frame, all portions of which are stationary. The " copy " is placed upon a convenient holder just above the keyboard, and above and behind

(*Continued on page 150.*)

SETTING TYPE BY MACHINERY, AS CONDUCTED AT THE NEW YORK "TRIBUNE" OFFICE.

Figure 8  Illustration from the first article about the Blower Linotype
*Source: Scientific American, 9 March 1889, p 1*

the use of movable type. It was published in January 1887 and carried the message about the Mergenthaler machine on the reverse of the title page. The *Tribune* carried no further information about Mergenthaler or the Linotype until 19 May 1889.

# ❦ THREE ❦

## *Large scale production and ensuing contention*

In the late nineteenth century, before radio and television, the newspaper was the fastest medium for disseminating news. Newspaper production was probably the most time-critical activity known and every edition was a cliff-hanger that left the staff drained when the paper was finally 'put to bed'. This was a cause of contention between Mergenthaler and the syndicate; they worked to tight deadlines and looked for quick results; he refused to panic, showed no sense of urgency and apparently wasted time over trivia. Reid was probably infuriated by his complacent attitude when he said that the working of the first machine, while not perfect, was very encouraging. He was satisfied that it would reach expectations before the first dozen machines were put into service and on 11 September 1886 wrote: 'The wearing out of matrices did not disappoint me in the least as I knew it from my former experience. The brass matrices will do better. Of course there are still a number of little defects; they are however of a nature which time and experience will overcome.'

### 1  *Slipping schedules and teething troubles*

Mergenthaler and the syndicate were often at cross purposes because he still considered the machine to be experimental while they thought it was fully developed. The Board treated his development plans and objectives as firm commitments. Further, he either gave, or agreed to, over optimistic delivery dates. Reid realised that Mergenthaler was thin-skinned and liable to take umbrage when criticised and, at least in the early days, tried not to upset him too much. When Mergenthaler wanted to keep machines back in the works to train his employees Reid compromised by sending men from New York to get operating experience when the Baltimore machines were idle. To avoid friction he often contacted Mergenthaler indirectly through: E. Lambert, his assistant, on routine matters; Milton White Johnson, the clerk at Baltimore, on production details; and Hine, on policy. Reid often confided in Hine who was on good terms with the inventor. In autumn 1886 he made the tolerant remark: 'I am greatly interested as well as amused at your news of the latest improvements – concerning which I will try to preserve a discreet silence,' but also showed exasperation when he asked: 'Do you know of any one promise that has ever been made in our Baltimore shop that has been kept?'

On 22 July 1886, some three weeks after the installation of the first machine, Mergenthaler invited Thompson, Reid's machinist, to discuss the arrangement of the keyboard and the matrix question. The original layout of the Blower keyboard is not known, but diagrams submitted with early Mergenthaler patents suggest that it may have been alphabetical. About 50 years later Charles Letsch recalled that when operators complained that some of the most used keys were too far apart, Mergenthaler tore out a two column story from the *New York Sun* and asked him to count the frequency of each character. He used this data to modify the keyboard to the standard 'etaoin/shrdlu' layout that remained largely unaltered on all later models. This made it simpler to operate the machine and, by putting the most frequently used characters in the longest tubes to the left of the magazine, ensured that they were among the first to be distributed. Every Blower machine was restricted to one size of type which could only be changed by replacing all 107 matrix tubes; machines were identified by type size, such as minion or nonpareil, see 'type sizes' in glossary.

On 3 August 1886, soon after the first machine was installed, but before testing it thoroughly, the Board ordered 100 machines. Then shortly before the first 12 were completed they put in an order for a second 100 on 15 April 1887. Mergenthaler protested vehemently that it was unwise to order 200 machines costing $260,000 to make, based on the experience of only 12 machines, but the Board replied that they were quite satisfied with the current model. Although those machines worked fairly well there were many teething troubles and they needed a lot of attention. It was evident that they should be improved, but the Board, anxious for quick returns on investment, refused to allow any radical changes to the design.

On 9 September 1886, Reid asked Mergenthaler about delivery of the minion machine and if it could be packed in separate boxes, taken up by elevator and assembled in the composing room. He also asked about progress in cutting steel punches for nonpareil and agate (a small type face used to set advertisements).

He was very anxious to get a full font of agate early; and unless they were reasonably sure to have those punches within a month would try to send another man for that special work.

Mergenthaler replied on 11 September 1886:

> I am glad to say that the first minion machine is far enough advanced to commence a test by Monday noon. The keyboard is considerably improved and I expect that in consequence thereof this machine will be capable of doing twice as much work as the first one. The flying out of matrices while running over the track seems to be entirely cured, but there still remains the possibility of one occasionally flying out when it reaches the counter although it is much improved in this respect also . . .

He wanted to keep the minion machine in Baltimore until the three others were built, to make the workmen familiar with its details. The machine could be dismantled and taken up to the composing room by elevator. Mr West had completed the nonpareil steel punches and Mergenthaler was sure that the agate punches would be ready when needed. He would like another type cutter if they could get someone less pretentious than West.

By 16 October Reid had become very concerned about money, deliveries and the threat of competition and asked Hine if he had heard about Talbot Lanston's machine [a prototype of the Monotype which was not produced commercially for several years]. He continued: 'The rate at which we are paying out money now bears no sort of proportion to the results we are reaching, and we must consider seriously the question how long we can wisely or with justice to the stockholders continue it.'

Despite complaints about rising costs and slow deliveries, Reid tried to avoid antagonising Mergenthaler and, in general, his letters to him were less blunt than those to his fellow directors. Before the end of 1886 Reid complained: 'The delays are proving interminable.' He wanted a shop in New York with a machinist who was not involved with new inventions, but dedicated to the quickest and cheapest production of inventions already made. It was not possible to be both an inventor and an effective plant supervisor, but he did not want to discourage Mergenthaler and wrote to Hine: 'I see absolutely no security for us save in a big shop here and a small experimental shop for him wherever he wishes to stay.' Reid complained strongly that the minion machine was worse than any of the others, mainly because of defective matrices. He wanted to send back the brevier machine for repair. This was the first model that had been 'working beautifully' on 2 July 1886.

## 2  *Pressure to produce large numbers of Linotypes*

Mergenthaler was not equipped for full scale production when the Board ordered 200 machines. He had been running a development shop with a few young assistants and suddenly had to organise and manage large scale production of a new product. This would have been a major task for an experienced works manager with no other responsibilities, but he had to carry out those duties while working as inventor of the machine. He had to recruit and train new staff, which increased the work force from about 40 to 160, expand the Camden Street factory, and rent a building on Preston Street where more than 100 men made matrices and assembled machines. He designed many special tools to produce and test machine components. He also gave contracts to firms in Baltimore and New York to make some of the larger parts. Mergenthaler described it as a Herculean task, and in an interview given to a Canadian paper some six years after his death, his widow said that he often did not return from the factory until late, yet set off for work again at 4 am. Obviously he overworked himself, but the root of the problem was that he did not know how to delegate. He should have trained others to supervise the routine work so that he could concentrate on the creative side of the business.

Production problems multiplied because contractors delivered late and supplied defective parts. Mergenthaler showed his lack of management skills when he said that he faced two dilemmas: the delay and cost for the reworking of defective parts or an even longer delay to wait for replacements. He did not mention whether he withheld part payment to compensate for the failure to fulfil contracts and apparently did not ask the Board to help him to get a more satisfactory service. Even if his pride stopped him from contacting the syndicate, he could have sought advice from a sympathetic member of the Washington group such as Hine.

Mergenthaler calculated that matrices should cost no more than six cents each for the Linotype to be economically practicable. He approached Mr Ryan, a well known Baltimore type founder, hoping to obtain outside help, but Ryan was dismissive. He showed a simple matrix that cost him between 50 cents and a dollar and said that the inventor could make a fortune if he could supply type founders with such matrices at 50 cents each, but Mergenthaler was more interested in solving the problem than in making money. He spent several months designing and making about 30 machines to produce serviceable matrices economically on a commercial scale. Reid complained that the matrices in the first 12 machines were so bad that he would stop the machines at his office unless improved matrices were supplied quickly. Early problems included: untrue matrix moulds; warped matrices that constantly blew out of the channel; metal that squirted between matrices and spoiled the cast; and matrix moulds made from electrotypes of movable type that were pulled off the matrices after repeated casting and made useless. Further, the slugs produced by early machines contracted so much on cooling that they would not print properly when mixed with movable type.

Attempts to cast matrices failed because no metal was simultaneously soft enough to take a good mould and hard enough to be cast from without melting. Steel dies punched into brass blanks would spread the mould, which interfered with justification and prevented impressions of uniform depth. Eventually satisfactory matrices were made with specially hardened brass. John Miller recalled that: 'for two years after that first machine came, the three Burr machines were used to help out with the typesetting,' which justified the policy not to change the dress of the *Tribune* when the first Linotypes were installed.

A major problem at that time was that hand engraved masters costing $5 each varied slightly in size and shape which would lead to unacceptable variations in the appearance of the typeface across the page. Mergenthaler started to design an engraving machine to mechanise punch cutting but, before he was able to complete this work, Dodge had acquired the Benton-Waldo engraver from the Northwestern Type Foundry in Milwaukee. The Foundry had tried to sell Reid their so-called 'self spacing' type (an early unit system). In the course of conversation Waldo told Reid, who in turn told Dodge, about their engraving machine which cut accurate copies of type in *soft* type metal from which to grow electrotype matrices. Dodge persuaded Benton to try cutting a steel punch on his machine; it worked, and a leasing agreement was arranged.

In the biography Mergenthaler commented wryly that it was a pity that the Benton-Waldo engraver had not come on to the market earlier. He noted that the first machine was not delivered until after his resignation in the Spring of 1888. However, he was involved in preliminary negotiations with the Milwaukee company. On 10 January 1888, Reid wrote: 'Please send on to the above address at once at least 100 steel blanks of the proper size for cutting a long primer or a small pica face.'

## 3   *Concerns about production and management*

Manufacturing had started reasonably and the order for 12 machines had been filled by the spring of 1887, but none of the first lot of 100 had been produced. Mergenthaler noted that: 'Mr Reid's almost daily letters became what might be called one great and continuous jeremiad of complaints about slow results.' He was told that they all appreciated his efforts very much, but hundreds of thousands of dollars had been spent with only the first 12 machines to show for it, that they did not believe that he was the only person who could make the machines and other things that were not at all pleasant to hear. Yet, despite admitting that he had received those messages he was quite unable to appreciate their importance. As he had not received any royalties he probably felt that the Board had no cause for complaint. He ignored all adverse criticism but remembered all comments in praise of the machines.

Progress was very slow in the assembly shop despite Mergenthaler issuing written instructions and staying long hours with the men. Even by employing more highly skilled labour, paid between $3\frac{1}{2}$ and $4\frac{1}{2}$ dollars a day, he could not clear the bottleneck. The men insisted on their agreed wages regardless of their aptitude which was so variable that some assembled a machine for $40 while others made the same work cost $250. Mergenthaler, who always tried to reward dedicated men, showed great ingenuity by introducing a bonus system that paid the assembler a bonus of $10 for assembling a machine satisfactorily within a specified time. This type of piece work produced the desired result. Some men earned much higher wages, but saved the company money because of the increased speed of production. Workers who did not earn the bonus received regular wages. Mergenthaler then introduced contract work into the manufacturing department with a similar increase in productivity. By February 1888 over 50 machines had been delivered to members of the syndicate with the balance of the first 100 due the following month. Yet, regardless of his extra responsibilities, Mergenthaler remained on the salary he had been paid to manage the development shop.

As Mergenthaler came under increasing pressure, he had a real cause for complaint when members of the Board interfered with his management of the works. After Reid complained about the quality of the matrices he dismissed Richard Berger, the foreman of the matrix department, who had not followed orders, and replaced him with a man who would produce work of the required standard.

In the biography, Mergenthaler claimed that Reid established a special shop where Berger was employed, at increased wages, to make matrices in competition against the company, with the comment: 'This little side-show was kept going for years and never produced matrices which were used for more than a few days.' However, on 16 March 1887, Hine wrote to Reid that Clephane and he had told Berger that they would give him a place to work, the material for making twelve sets of matrices of twelve hundred each and pay him fifty dollars per set if they were satisfactory. Evidently Mergenthaler was aggrieved and Reid tried to placate him with a letter on 11 April implying that they were trying to take the strain off him. Reid had feared that Mergenthaler would regard the engagement of Berger as a reflection on himself but there was no real reason for him to feel hurt. Berger had said he could make much better matrices in a very short time and backed his promises with his money, putting up $1,250 which would be forfeit if he were not successful. Even so, it must have been galling for the inventor to receive a letter dated 7 January 1888, from Reid's assistant, Lambert, that included the sentences: 'The sample marking has been sent to Mr Berger with the request that he have it cut immediately. I have also told him to see that the punches in future are made longer.'

In another case, the clerk M. W. Johnson, brother of E. Kurtz Johnson, a major stockholder, was often drunk and neglected his work. Mergenthaler, 'politely told him that he had to do better in future,' and, according to the biography, was astonished and disgusted when the man was given a rise in pay and made an independent employee of the Board with responsibility for writing weekly progress reports. Mergenthaler only questioned the man's competence to write those reports. Johnson had been one of Reid's regular contacts before the schism and did not write spitefully about Mergenthaler. In fact, Berger and Johnson had worked with Mergenthaler before the syndicate took over and were named as witnesses on some of his early patents.

Mergenthaler waited until machines were working in syndicate offices before raising the question of royalty payments. In an aside to a progress report written on 14 November 1887 he reminded Reid of his agreement with the National Typographic Company. The latest machines could earn revenue and he asked for an account of the outstanding royalties. Lambert replied for Reid on 17 November that he knew nothing about the contract with the National

Typographic Company for a one-tenth royalty and would lay the matter before the Executive Committee.

Reid's letters were not all hostile. On 3 December 1887, he wrote: 'The machines lately received are giving great satisfaction. Our chief difficulty now is in training operators fast enough, and this has all along been more serious than we expected.' On 12 January 1888, he wrote: 'Mr Haldeman complains that the working of the nonpareil machines he has already received are [so] unsatisfactory as to discourage him in their use. This is probably the fault of the operators; and you will no doubt be able to remove any bad impression he may have gained with the five now ordered . . .' a clear indication that, in this case, he considered the blame to lie with the user rather than the inventor.

## 4  Reid's 1888 report and the start of strife

In the biography Mergenthaler claimed that Reid told an audience of at least 100 stockholders at the annual meeting on 21 January 1888, that: 'so far the Linotype machine had . . . brought nothing but trouble, loss and disappointment to those using it,' and that 'syndicate members were ready to discontinue use of the machines at any moment.' He described this as 'like a thunderbolt out of a clear sky'; it was too much for him and he contradicted Reid: 'in open meeting, in the interest of truth and fairness.' However, nowhere in the report does it say that syndicate members were ready to discontinue using the machines.

Mergenthaler had noted the 'continuous jeremiad of complaints about slow results' and comments that he found hurtful. Reid and his foreman encouraged him when things went well and he admitted that the machines were not perfect but he seemed to be unaware of the frustration of people trying to produce a newspaper with unreliable equipment. He probably over reacted to Reid's report and as a German with English as a second language may have misunderstood the idiom, may have taken items out of context, or Reid may have made an unreported aside at the meeting. The inventor often claimed how he had politely, but firmly, disagreed with others and presumably thought that once he had stated his position, the matter was settled. However, his own account of his outburst at the annual meeting showed that it was unwise, intemperate and extremely rude. He seemed unable to realise that such behaviour could cause lasting rancour.

To get a more balanced view of the situation it is necessary to consider parts of Reid's report. He stated that the first 12 machines had finally been completed in the summer of 1887 and that considerable sums had been spent to keep them in running order. Several had been laid aside as almost worthless until reconstructed and most of those still in use were only available for training beginners. Reid gave a list of where machines had been delivered and were to be delivered, and stated: 'No one of these machines has yet earned a dollar for the company or for anybody else. So far from proving a saving, the introduction has thus far been a source of materially increased expense to every newspaper that has really attempted to introduce them.' He noted that one syndicate member had been in dispute with the union over rates of pay when the machines were installed and another faced a strike of the entire force. He had expected the *Tribune* to make a profit because it was non-union.

Reid listed his composing room expenses for the years 1884–5 when there were no Linotypes and 1886–7 when there were Linotypes:

| 1884 | $73,176.26, | 1886 | $86,009.71, |
| 1885 | $75,070.16, | 1887 | $83,475.37. |

These figures are not as sinister as Reid implied. Rising costs in 1885–6 suggested that a shop without Linotypes would have cost about $81,000 in 1888. In 1886–7 the cost of running the composing room with Linotypes was falling and should also have been about $81,000 in 1888. Therefore both systems should have cost about the same in 1888 and from then the composing room with Linotypes should have been more economical.

Reid remarked that 'the damage done by the crude results attending the work of the first machines is beyond the belief of outsiders,' noted that he expected the position to improve with experience and commented: 'In this, as in everything else, it is the pioneers who suffer.' Far from condemning the machine out of hand he said that problems of producing good matrices would be solved with the Benton-Waldo equipment and that the quality of slugs was improving. He complained about the difficulty of changing the measure and the fact that each machine could only carry one font of type. Further, he said that there was no adequate device for the use of italics or small capitals. In 1936, Martin Q. Good, an early operator who joined the composing room staff in June 1888 said: 'At that time all of our tabular matter was still set by hand, and when small caps or italics were called for in a line the slug was cut and hand types were inserted.'

However, Reid did attack Mergenthaler, but so subtly that the inventor did not even notice. The sardonic phrases are in italics. In comments about general improvements he wrote: 'The machine has now reached the stage when *a hundred minds instead of one* can be readily and profitably directed to its improvement.' In remarks about production costs he wrote: 'while *the man at the head of the manufactory* was engaged in necessary experiments, the work suffered from lack of superintendence.' Then he noted that the company had set up a small shop in Brooklyn, under Mr Berger, formerly Baltimore foreman, which has done some excellent work on matrices – this was 'the little side show' previously mentioned.

The report suggested that typists were likely to make the most efficient operators; that no machines should be let out to other parties until the syndicate was fully supplied (hardly the sentiments of one opposed to the machine) and implied that it would be unwise to use the machine in small offices because they required 'care from a practical machinist' and 'a few failures to utilise single machines in small offices would do great harm.' Finally, the word 'Linotype' was not used in the report – it only appeared in the imprint on the cover and title page as NEW YORK TRIBUNE LINOTYPES and in the 'Terms for Machines' at the end of the report.

Other than to Mergenthaler who objected to adverse criticism, even when justified, Reid's comments about the Linotype were fair. He had reservations about its performance, but expressed long term optimism and kept the machines in his composing room. He advised friends that the stock was highly speculative and took a calculated risk by buying all the stock that he could obtain at a reasonable price.

Under its founder Horace Greeley, the *New York Tribune* had a fine reputation with working class readers as a radical journal with a social conscience. When Reid took over, the paper became decidedly reactionary and he was attacked, as shown by the contemporary lampoon in figure 9. It is obvious that Reid, an autocrat who threw the union out of his newspaper following a strike in 1877, would not tolerate what he would regard as treachery or impertinence by the equally stubborn inventor.

## 5   *Mergenthaler's resignation*

The climate changed immediately after the 1888 stockholder's meeting, but Mergenthaler did not seem to have noticed. Reid dropped all pretence of paternalistic tolerance and started to harass him in petty bureaucratic ways. In this he easily ran rings round the inventor who was plain spoken rather than devious. On 29 January 1888, Mergenthaler wrote to thank Reid for the indulgence shown in the matter of the loan and asked to be paid for his tools and shop equipment. He wanted to repay the loan as soon as the Company settled his bill and could not see why it was not met like all other bills incurred to increase the facilities of the factory. Reid replied brusquely next day that the note sent for the interest was defective in form and enclosed a correctly made out blank. He referred questions about Mergenthaler's royalties and payment for his tools to the Board of Directors; a procedural ploy to delay payment.

At this time there were problems with cash flow and Reid complained about running costs. Mergenthaler, having originally been told to produce machines as quickly as possible, was now urged to economise. On 25 February he was told that it was vital to reduce the pay roll and to 'weed out the least useful men.' Reid ended: ' My earnest advice to you is to endeavour to reduce the expenditures in all these particulars to the most economical basis.' Mergenthaler, who must have taken all words at face value; was quite unable to realise that Reid's 'earnest advice' was an order couched in polite terms. He complained that this policy would lead to the loss of a highly intelligent workforce that had taken years to put together and on 28 February decided to assert his authority when he replied: 'I will continue to fill the position to which I was appointed myself; respectfully declining to accept the kind advice given.' Reid was in no mood to take any lip from Mergenthaler and wrote to him on 9 March in no uncertain terms with a set of directions that were to be obeyed.

Mergenthaler resigned on 15 March 1888. He wrote that he had been reluctant to become workshop manager and had only accepted because he was supposed to be in sole charge. This responsibility had been usurped by the Board of the Mergenthaler Printing Company. He wished the company well and would help it in every way possible. He would never again work for a fixed salary but was prepared to accept contract work. His ambition was to become head of one of the largest commercial establishments in Baltimore and he regarded his ability as an inventor as a means to that end.

The account of correspondence leading to Mergenthaler's resignation given in the biography was copied from Reid's 1889 report. [It is not known how this unique hand-written and unprinted, though marked up, report came to be among Mergenthaler family papers.]

Mergenthaler was told to do no work on any machines beyond the 200 ordered. He was anxious to leave as soon as possible and, likening himself to the captain of a doomed ship, wrote: 'All he wanted was to get off and it

# LARGE SCALE PRODUCTION AND ENSUING CONTENTION 55

Figure 9
A lampoon of Whitelaw Reid

mattered little to him whether he was honourably relieved, discharged or thrown overboard.' With his resignation he asked to be paid for his tools (about $6,000) and his ten per cent royalty. Reid replied that the claims would be laid before the Board, but this was a delaying tactic because, for all practical purposes, he was the Board. (Stone left the syndicate in 1888 having sold his interest in the *Chicago News* to his partner Lawson for $350,000 plus an annual payment of $10,000 for ten years, provided that he did not enter into competition in Chicago.)

Reid accepted Mergenthaler's resignation on 4 April, noted that he would be considered equally for outside work, and asked him to turn over the Camden Street factory to M. W. Johnson and the Preston Street factory to R. B. Bartlett. When Mergenthaler asked to be paid for his tools before handing over, Reid said the tools had nothing to do with his resignation and demanded immediate surrender of the factories, an order that was promptly obeyed. Thus far Mergenthaler had considered this matter to be a mere business disagreement but finally decided that Reid was punishing him for his dissenting views. He feared that Reid would try to avoid paying his claims for tools and royalty and engaged Mr Chas Marshall of the Baltimore bar to represent him.

There were mixed reactions to Mergenthaler's resignation. Hutchins tried to keep everybody calm. Smith was very angry and wanted formal assurance that Mergenthaler would not be re-engaged. Hutchins thought it unwise to pass resolutions against employing Mergenthaler or of putting anything on record to exasperate him. He had resigned but by careful sensible management one should be able to get the full benefit of his great inventive ability.

Mergenthaler believed that he was the only person who could make a Linotype and thought that he had the company at his mercy. He could not see that a competent engineer would be able to duplicate and possibly enhance the Linotype using the example of machines installed in syndicate offices. Hutchins reported that Mergenthaler was willing to enter into a contract for making machines and sent Reid a copy of his tentative proposals. He had talked plainly to Mergenthaler and thought that he could be dealt with from Washington to the profit of the company, but not from New York. There may be legal problems if he were dealt with too bluntly.

Mergenthaler's outline proposals, which were not binding, included making Linotype machines at a fixed price, to be determined later, subject to the company selling him their shops as they now stood at a valuation of about two thirds of their cost for all tools now used to make the machines; that regular payments be made as the work progressed and that no more than one-sixth of the total price be retained until the final acceptance of each machine. He should expect a contract for at least 400 machines. The payment for the shop could be effected by the one-tenth royalty due on the cost of the machines or he *would be willing to sell out such Royalty at a liberal price to the Company or its members, thereby enabling him to pay cash for the shop.* [This was the only time that Mergenthaler was reported as being prepared to sell his entitlement to royalty.] Hine thought it unwise to sell him the tools at any price, but thought it best to give him contracts at fair rates for some parts of the machines.

However, on 9 April Mergenthaler wrote to Hutchins: 'In consequence of a letter received today from Whitelaw Reid I withdraw herewith my offer in regard to building these machines under contract. The letter in question shows plainly that the disagreement between us has taken the shape of a personal ill will and that it is his purpose to make it troublesome to me wherever he has the opportunity. Under the existing circumstances it would be madness to attempt it.'

In two letters written at this time Reid told Hutchins: 'No employment could be given to Mr Mergenthaler in the shops and no authority of his recognised there, pending the completion of the present machines; but if he wished to experiment and make improvements, he would be provided with a place and tools.' The letter which had made Mergenthaler withdraw his proposal, was one objecting to his effort to hold on after resigning. Reid ended: 'There is no use fooling with him now. After he has got off his high horse, he will be valuable again.' He did not want to alienate Mergenthaler completely, he just wanted him to climb down!

A press release in the *Morning Herald* of 11 April 1888 announced: 'Mr Ottmar Mergenthaler, inventor of the typesetting machine named after him, and during the past four years manager of the Mergenthaler Printing Company, has resigned, owing to differences with the directories. Mr Mergenthaler says that the directors desired to move the works of the Company to New York. He preferred to resign rather than have that done.'

In other words Mergenthaler colluded with members of the Board to avoid revealing the real reason for his resignation. However, Reid still referred to Linotypes as 'Mergenthaler machines' after the 1888 report.

Other members of the enterprise were obviously unaware of the bitter exchange of letters between Reid and Mergenthaler. Clephane apparently thought that Mergenthaler had taken umbrage over a minor matter when he wrote: 'In the meantime, the company insisting on the removal of the shops from Baltimore to New York, Mr Mergenthaler took offence, sold his stock, and gave up all interest in the enterprise.'

## 6   *Production at Company factories after Mergenthaler resigned*

The dispute between Reid and Mergenthaler triggered several events with far reaching consequences. The Executive Committee asked Hine to stay in Baltimore 'under pay' to complete the second hundred machines and invited several firms, including the Colt Arms Co, to bid for contracts to manufacture the Linotype. Reid, anxious to consolidate his position by praising those he supported, claimed that Berger was doing the best work on the machines in the little shop in Brooklyn. He had perfected a method that used type metal instead of soft metal to produce better linotypes with a sharper face. This may have been an early attempt to produce slugs with the durability of movable type.

In 1924, the Australian printing historian Henry Lewis Bullen, recalled that in May 1888 he was due to sail to Europe from New York at 5.30 am and had been with friends until 2 am. It was too late to return to the hotel and too early for the boat, so he went to the *New York Tribune* and saw the third version of the Blower Linotype which produced slugs as good as those of contemporary machines. He was so impressed that he took 60 slugs with his name in 6 point capitals to show to people in Europe and paid an operator $35 to send him a monthly letter for a year giving news of the machine. The letters were proofs pulled from Linotype setting. Within a couple of months, requests for machines arrived from Australia.

Hutchins wrote to Reid that Hine and he had discussed the situation at the factories. Hine and Mr Julien P. Friez, Superintendent of the Camden Street works, had studied the machines in detail and concluded that under Friez's management they would cost at least $700 to make. [Mergenthaler estimated $1,300 and the Colt factory quoted $2,200.] Friez was abroad because his father was seriously ill, but had agreed to return if needed urgently. Hutchins thought that he should return to superintend the shops. Hine thought that Berger should be employed in a small shop and as an inventor because he was not as competent as Friez to manage a large shop.

People who enquired about the Linotype were told that they were not for sale. They were only available on lease for a non-returnable deposit of $1,000 with a royalty of 10 cents per thousand ems of printed matter. One prospective user was told that if a machine was returned the difference between its depreciated value and the cost of refurbishing it, if any, would be refunded. Apparently there were no takers at those prices.

Members of the syndicate voted themselves special rights. They would be supplied before any one else and have exclusive use of the machines in their respective cities. Thus the *Tribune* had the monopoly in New York City and was the reason for Hutchins refusing to supply Linotypes to his former colleague, Joseph Pulitzer. The only user outside the syndicate was the *Providence Journal* which had acquired a dozen machines through Abner Greenleaf, one of the Washington group. Syndicate members also cut their royalty on production from 10 cents per thousand ems to 1 mill (one-tenth of a cent) per thousand, with free maintenance and upgrades, but charged an annual rental of $500 with a 20% discount to early users.

On 24 April 1888 the Board engaged Mr Charles H. Davids, formerly superintendent of Edison's workshop, on a temporary basis, and appointed him General Manager of Manufacturing on 14 May at $4,000 per annum; $1,000 more than the inventor was paid. Mergenthaler described Davids as an elderly gentleman, well past 50, a very fluent speaker, apparently full of energy, but with no experience of Linotypes. In fact Davids came highly recommended; he was a mechanical engineer who had patented a machine for making stereotype matrices in 1883. He found conditions chaotic in Baltimore with two factions: 'Mergenthaler men' and 'anti-Mergenthaler men'. Some of the foremen were discharging each other without authority. The men had learned of the intended move to Brooklyn and production had suffered. Davids did not know how they had found out although it had been part of Mergenthaler's resignation statement in the *Morning Herald* on 11 April.

Davids seems to have been an abrasive character who tended to clash with everyone. He antagonised the workforce at Baltimore by finding a loop-hole in a cancellation clause and used it to cut the payments to teams working on contract. Mergenthaler criticised Reid for accepting 'this deplorable deed without a word of censure' but Reid had complained about the sums paid for contract work, particularly to Friez, who resigned in May 1888; presumably his payments had been cut. Davids discharged all the workmen with the exception of a few foremen and said that they might apply for employment at the Brooklyn factory when it was ready.

The main factory was closed on 1 July and the matrix shop on 7 July. Davids also disagreed with Berger and Johnson. In July, Reid advised him to handle

Berger with a light rein, but left the decision to Davids. In September, Davids sacked Johnson who claimed that he was appointed by the Board and could only be dismissed by them; but the Board supported Davids. Reid gave Johnson a reference and noted that he did not mention the complaints about his occasional tendency to drink, he told the prospective employers that Johnson had worked for the Mergenthaler Printing Company until a few weeks ago when, owing to some difficulty with the new General Manager, his resignation was requested. Reid said that Johnson was energetic, attentive, a capable book-keeper, desirous to promote the interests of his employers and ended: 'He is a brother of E. Kurtz Johnson, President of the Citizens National Bank of Washington, and head of a coal company.' [The last comment may well have been the kiss of death; if his affluent brother would not help him, why should anyone else?]

By mid-October, the Washington group had become disillusioned with both Reid and Davids. On 17 October Reid told Davids that the Directorate was opposed to selling him shares and continued: 'Do not forget that at present all the acts, both of the shops and the executive office, are subject to a very jealous and hostile scrutiny.' Mergenthaler had shown that Davids's payroll estimates implied that he would immediately make 15 machines per week for no more than $133 each in wages and that within a year he would make 30 machines per week for no more than $100 each. Reid wrote on 24 October that members of the Washington group had been to see him that day, much disturbed by Davids's sanguine calculations. Reid assured them of his satisfaction with Davids's work in the shops and confidence in his management and hoped they were satisfied. He said that they might visit the shops next day and asked Davids to show them through the works. He also offered Davids a few shares at par [below market price and against the wishes of the other directors]. Even so, Reid censured Davids for exceeding the estimates for fitting out the factory, paying too much for punch cutting, and authorising payments that were not properly processed. Davids complained that Garvin supplied unsatisfactory castings but, unlike the inventor, referred the matter to Reid who followed it up. He also complained that production of the small pica machines had been delayed because Benton, Waldo and Company had supplied defective punches and many had broken in use.

Reid had cash flow problems throughout his term of office. On 2 May 1888, he wrote to members of the syndicate that less than $2,000 remained of $32,000 paid in. There had been a heavy bill from Dodge for home and foreign patents and more money was needed urgently. Some six weeks later, Reid complained to Hutchins about late payments and asked him to tell the Washington group, who supported Mergenthaler's claims, that if they were so anxious to pay money out, they should not be reluctant to pay their share. Reid also told Davids not to let machines out without written orders from his office; the *Washington Post*, for example, was in arrears and must pay up before receiving any more.

## 7   Mergenthaler on his own again

Reid procrastinated when Mergenthaler instructed Marshall, his lawyer, to pursue his claims against the National Typographic Company. On 20 April 1888 Reid replied that the claim contained errors of fact and he wanted to wait until the meeting of the Board of the Mergenthaler Printing Company, expected within a fortnight, for the directors to decide. He claimed that Mergenthaler did not have to turn over his tools when his resignation was accepted, he was ordered to turn over the shops. The claim was against the National Typographic Company; its relations with the Mergenthaler Company had not been settled. It was an alternative claim and, even if the company accepted responsibility, there was no evidence that the options named in the contract had ever been exercised. The Company wanted to treat the inventor justly and liberally and shared his desires that questions between them should be settled amicably, although they did not see that resorting to lawyers was the best method of attaining that purpose.

Mergenthaler was anxious to start a new business and bought the burnt out Walker horseshoe factory at the corner of Clagett and Allen streets in Locust Point, not far from where he had disembarked 16 years earlier, see figure 10. These were substantial premises, about 280 feet long by up to 120 feet wide, with a private railway siding. Early in May 1888, having no income and with his claims against the company unresolved, he sold half his stock in the two companies to clear his debt of some $9,000 with Reid and make a down payment on the new premises. The rebuilding costs were about $7,000 and, as he needed cash to buy new equipment, he disposed of the rest of his stock in lots of 100 shares – but their value was declining so much that the last lot did not realise the assessments paid on it. It was said that he sold his shares for $40,000 as though this showed affluence, but as he originally held stock with a face value of $50,000 in the two companies and paid interest on the loan he was a net loser on the transactions.

On 16 May, Smith wrote to Mergenthaler and next day Reid informed Marshall that the Board wanted to settle the inventor's claim for tools and consider his claim for royalties. Mergenthaler replied on 19 May that he assumed that the executive committee had authority to act for the National Typographic Company and noted that the long delay had caused him much inconvenience and loss. He was pleased to hear that they intended to make an equitable settlement for his tools and machinery and he was ready to meet the committee any day during the coming week to discuss the royalty claim. He hurried to New York to meet the company secretary in a room in one of the office buildings. Almost immediately Reid entered and the secretary left the two men on their own – the very situation that Hutchins had advised against.

Reid wanted Mergenthaler to release the company from having to pay royalties. When he threatened to go to court, Reid retorted: 'The courts! We have nothing to fear from the courts, it is an unjust contract and it is an illegal contract and courts are not very much given to enforce such contracts as that one.' He said that the Board wanted to reach an equitable settlement which meant paying nothing on the syndicate's machines. At this point Mergenthaler walked out, 'more than ever wondering about Mr Reid's queer ways of assisting him in building castles on the Rhine.'

Reid reported the same meeting in a memo to Smith of 31 May, and showed that despite his apparent confidence in parrying the inventor's claims, he was worried about the company's legal position. Mergenthaler stuck tenaciously to the claim for 10% of the cost of making the machines, estimated at $1,300 each. This would cost about $15,000 immediately and not much less than $26,000 on the 200 machines. The contract with the National Typographic Company did not bind them, but even if it did, the royalty applied only to machines that brought in a revenue. They would be glad to hear any suggestion for an adjustment, but Mergenthaler said that he did not see how there could be any adjustment if machines bringing in no revenue were excluded from the contract and left, apparently to consult his lawyer. Reid had thought it best to delay settlement on the tools, but would not object if Mergenthaler wanted to take them. The company was waiting for Davids to say if they wanted to buy them on the agreement made with the National Typographic Company. Mergenthaler thought the company was trifling with him; he had been invited to meet the committee and had met only Reid. The memo ended: 'I think we must really try to have a full meeting of the committee if we should meet him again. My fear is that he will bring suit. On the other hand my impression is that it will be cheaper to stand the suit than to settle with him on his present terms and acknowledge the contract as binding on the Mergenthaler Company.'

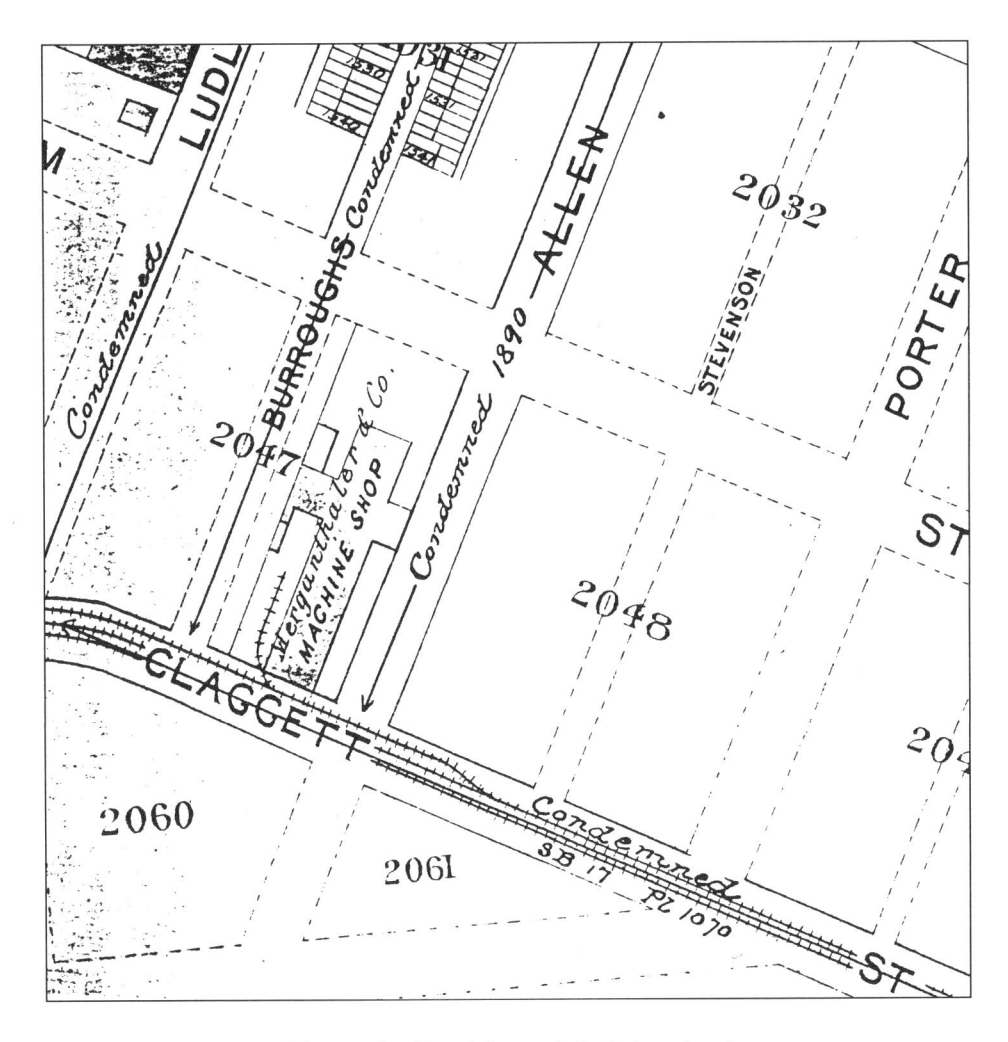

Figure 10   Top: Mergenthaler's letterhead
Below: Portion of map showing his works: scale 200 ft to the inch

Reid wired Davids on 31 May: 'If Mergenthaler wants his tools let him have them, don't give him any chance to say that we are trying to stop him reclaiming them.' Davids apparently ignored this order and Mergenthaler wrote to Reid on 1 June: 'I herewith ask you to either make an immediate settlement of the bill presented to you as representing the value of my tools, machinery, material and fixtures or else to deliver them to me within a week from date. Such delivery to be under the provisions of my Contract with the National Typographic Company.' To which Reid replied next day that the inventor should provide an inventory of tools and pick them up.

No tools were handed over so, on 10 June, Mergenthaler told Smith that when he called on Davids, the superintendent said that he had no authority to act in the matter. Later Davids sent word to see him on Monday [15 June] but did not seem to realise the difficulty of identifying the tools. The inventor needed cash to build his factory and the interest on his tools, valued at $6,343.40, for $3\frac{1}{2}$ years was over $1,400. The next day he informed Smith that when he went to the Camden Street works, Davids said he was ready to proceed and had orders to deal with him liberally.

When asked for an inventory, Mergenthaler showed his books and bills, but Davids said that it was no inventory. He wanted a list of every tool brought into the shops. Mergenthaler pointed out that any inventory would be based on those bills, but Davids replied that his time was too valuable to waste looking at bills and books and said in a high toned way that things were now being done in a business like way. To settle things Davids's way was to ask for the impossible; it implied that as no inventory had been made at the time he had no claim and would get *nothing* for the four years of valuable service from those tools. Mergenthaler asked Smith to tell Davids to abandon technicalities and either settle the matter on its merits or accept the bill presented and remit the full amount. This would probably do more to restore harmony than could be done by any other means.

Mergenthaler wrote on 17 June that some tools had been returned, but there was a considerable shortage and Davids pretended that he had no authority to make an adjustment. He wanted to know where to send the bill for the shortage because he had to pay cash for replacements. Smith added a memo to Reid: 'I told him to make out a bill and send it to you.' However, Mergenthaler was still kept waiting and complained to Reid, on 21 June, that he should have had an early answer to his letter of 17 June about overdue interest and said that he should not have to wait any longer.

Reid wrote to Davids on 30 June about closing the Baltimore works and moving to New York within a week and sent a hand-written letter telling Davids to try to identify Mergenthaler's tools and return them while closing the factory. Some time later he told Davids to take all experimental machines to Brooklyn and preserve them carefully in case of future patent litigation.

Throughout this period Reid was desperate for cash to meet the costs of closing the Baltimore shops, and called for $2,000 each from members of the syndicate and $1,000 from people like Clephane. In the meantime, he sold the Baltimore works with all equipment and unfinished machines, which Mergenthaler thought to be worth at least $180,000, 'to one Chas R. Williams, of New York, for the ridiculously small sum of $20,000', and swore an oath that it was a *bona fide* sale made by order of the Board of Directors. [It was discovered later that Williams was related to Smith, the company secretary.] Reid then conveniently leased back this property from Williams in order to move it to Brooklyn to complete the unfinished Linotypes. When Mergenthaler heard about this move he claimed that the sale was merely a ruse to stop him from delaying the move of the factories to New York. His lawyer suggested that he issue an attachment to stop the property from leaving the state without a bond to cover his claim. Reid said that it was a genuine sale made to raise cash quickly, there being insufficient time to make an assessment on the $600,000 of uncalled subscriptions. The company bought the property back from Williams, for the same price, on 14 January 1889.

These months of persistent harassment and procrastination were trivial in comparison with Mergenthaler's grief when his second son, Julius, died on 2 July 1888 at the age of four. He told Reid of the tragedy in a letter of 7 July, apologising for the delay in acknowledging receipt of the balance of salary but, surprisingly, there is no record that Reid sent a message of sympathy, a normal courtesy at that time. He also pleaded for payment of his claim asking: 'Would it not be possible to have a settlement of the latter during the coming week? I am very much pressed for money and compelled to sell Stock at a sacrifice while the Co is owing me money which is (at least in part) overdue for years.'

On 16 July 1888, marked unanswered in Reid's correspondence books, Mergenthaler wrote that the settlement of his account having been delayed in a most unjustifiable way, he was compelled to apply for an attachment to prevent the property of the company being removed from the State. If his claim had not been settled by Wednesday the 18th his attorney would apply for an order to set aside the sale to Chas R. Williams and sell the goods now held by the Sheriff.

Chas R. Williams would not be permitted to remove the property from the State before the claim was settled.

Lambert replied on 17 July: 'I regret that Mr Reid is away from the office today and unable to reply to you himself. As I understand the matter, however, he has no longer any authority over the property you refer to, the same having been sold to Mr Charles R. Williams.'

The company then moved the factory to new premises in Brooklyn, taking the partly finished machines and tools from the Baltimore plant. It retained the Mergenthaler name in its title, which led to false reports that Mergenthaler had moved his factory to Brooklyn.

Later that year Mergenthaler suffered a near fatal bout of pleurisy, lasting almost two months, that laid the foundation for the pulmonary tuberculosis from which he died. He recovered slowly but during his illness he conceived a completely new Linotype which overcame many problems of the Blower machine. Shortly after he resigned he had engaged Mr Alfred Peterson 'a very able draughtsman' to assist him. This was probably the work he liked best and at which he was most effective.

## 8  Financing the new machine

Mergenthaler's biggest fault was that he continually modified the current model and told his sponsors about the latest developments before they had been fully tested. Nevertheless, such was his credibility, expertise and charm that when he appealed to Clephane for funds to build an 'improved Linotype' towards the end of 1888, ten of the original Washington group each gave him an immediate loan of $200. Clephane claimed that Hine, Hutchins and others cheerfully subscribed $5,000 and the work began.

Hine telegraphed Reid on 22 September 1888, that because several men were sick, Mergenthaler's new machine would not be ready until about October first. There was no doubt that the new machine would be the future model. The design was completed towards the end of 1888 and the first machine was ready to be tested in 1889. This model, later known as the 'Square Base', see figure 11, provided the typical configuration of all subsequent Linotypes. However, Mergenthaler thought the chunky square base was too heavy and blamed the draughtsman for its design. [An inventor is responsible for the plans; the person who expects to take credit for success should accept responsibility for failure.]

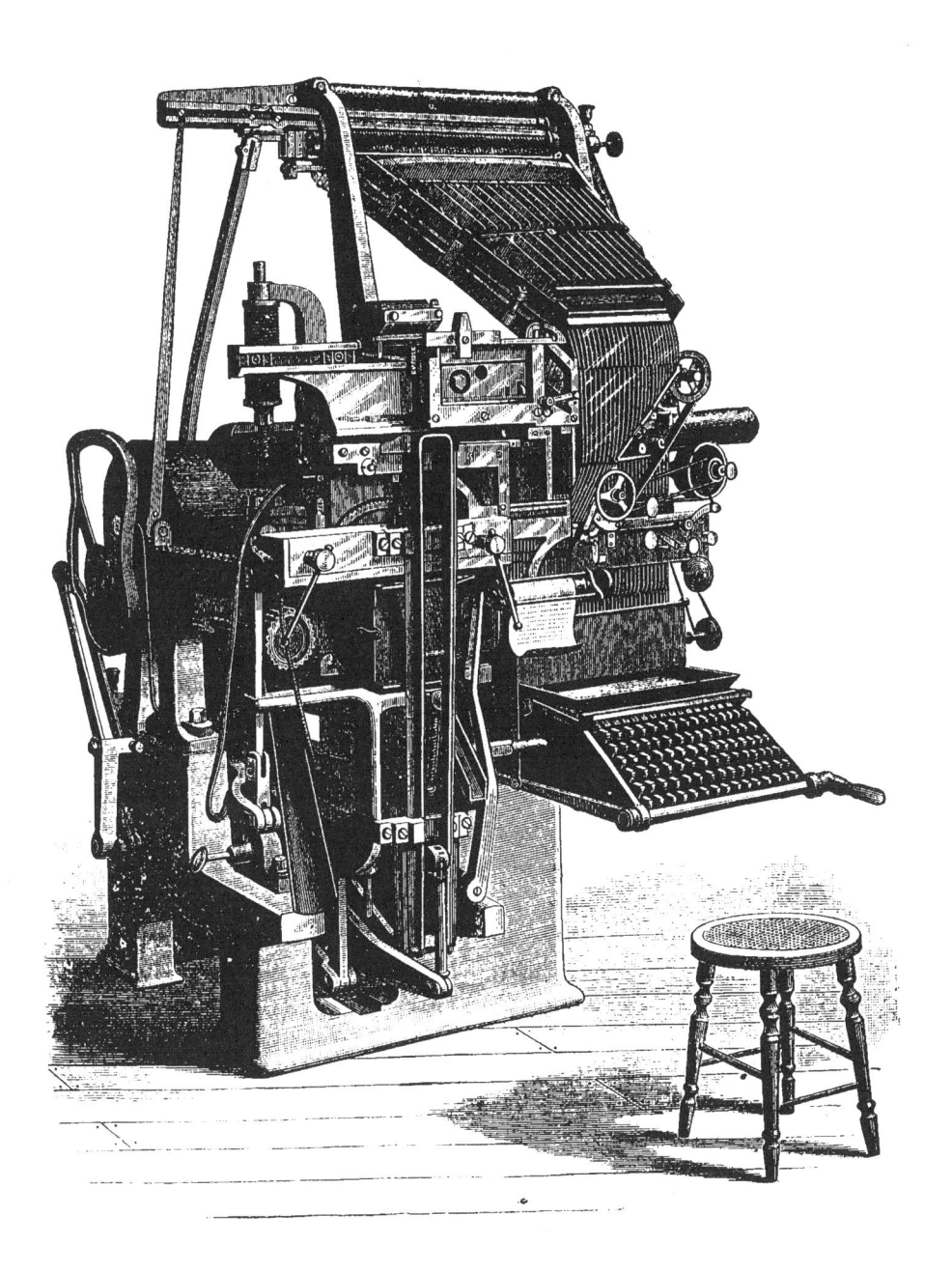

Figure 11  The 'Improved Linotype' known as the Square Base
*Source: The British and Colonial Printer and Stationer, 12 March 1891*

Although Mergenthaler did not contact Reid about the new machine, he was informed about it by Greenleaf and others. There was no reference to the new model in Reid's 1889 report, but he ordered Davids, who had finished only 40 machines by the beginning of the new year, to stop producing Blowers when the current orders for 200 were completed.

Mergenthaler was so dissatisfied with the square base that it was not illustrated in the biography. He immediately started to design a lighter version with the traditional star or claw base known as the Simplex; see figure 28, in Chapter 6. It was finished in February 1890 and exhibited in the Judge Building in New York by James Clephane and Abner Greenleaf.

In 1889, Fritz (Friedrich) Mergenthaler, Caroline Hahl's son, joined his half brother in Baltimore. Prior to this, he had been in the team which erected Linotypes in Belgium and France to protect the patent rights in those countries. Reid had become so concerned about the costs of this effort that Hutchins was sent to Paris to contact the team leader Knoop (a Baltimore machinist), and report back to the Board. Between 1 March and 24 August 1887 Reid often wrote to Mergenthaler about this activity. On 11 April he mentioned '$120 paid to your brother', but without details, and there were complaints that he was of little service. Reid complained about the costs of keeping people abroad, in particular that the men had been drinking wine; probably he did not realise that *vin ordinaire* was cheaper than soft drinks. Reid said that Mergenthaler's brother had written that they were erecting three machines instead of one in Paris [possibly at the Paris office of the *New York Herald* which employed operators with Blower experience] and had to pay two and a quarter times regular wages to get the work done quickly.

## 9  *Stilson Hutchins and selling the manufacturing rights abroad*

As the first newspaperman to recognise the potential of the Linotype and a contributor to funding the improved model Hutchins must have realised that it would sweep the market. He was one of the few people to discard Blowers from his composing room because of disputes with the union and maintenance problems. The *Journalist* reported, on 27 October 1888, that the contract between the Typographical Union and the *Washington Post* would expire on 7 November and that the managers of the paper would probably have some trouble renewing it. Nevertheless, Hutchins believed in the ultimate success of the Linotype, because he sold his newspaper and invested heavily in Linotype stock. When, probably during the summer of 1888, he met the Englishman Joseph Lawrence he devised a strategy for restoring the finances of the American company and disposing of surplus machines.

Lawrence probably only wanted Linotypes for his own use and may have considered starting an agency in Britain. Hutchins did not start a British branch, as reported by American sources; his brilliant plan was to sell the manufacturing rights to a British syndicate who would form their own company, take all the risks, and pay handsomely for the privilege.

Hutchins suggested that he sell the European manufacturing rights for $1,000,000 to earn $200,000 commission. This proposal may have been leaked because on 24 August 1888, Messrs Backus and Felton of New York City offered to negotiate the sale of the English patents. Reid replied that the company had not yet decided to sell its English patents but that a heavy stockholder had already proposed going to Europe to negotiate the sale in person. In December, Reid told syndicate member Rand, referring to the sale of foreign rights, that they were committed to Hutchins and would probably not have the right to begin negotiating with anyone else.

Reid made no secret of the fact that he was not enthusiastic about selling the overseas rights, but the majority of the stockholders seemed to be in favour and he did not want to be 'a dog in the manger'. He believed that nobody would pay large sums for the right to use machines until they were perfect. They were improving but still had many little defects. There were plans to set small caps in the same machine with the present font, and a skeleton machine for italics. There was a device to change the measure at will and another to change fonts in the same machine. When these worked they would have a merchantable article and could 'make a great fuss.'

The Executive Committee, and later the directors, took several months to consider Hutchins's proposal. On 6 November Reid wrote to Dodge that the companies had made no arrangement with Hutchins and, judging from talk at that day's meeting, the conditions would be seriously modified. The Board demanded much more than Hutchins proposed. In November 1888, he was authorised to sell, before 1 April next, either the English or the Continental rights for at least $1,500,000, or both sets of rights for at least $2,500,000; payment to be either two-thirds or three-fourths in cash with the balance in stock. Hutchins to be paid $250,000 for selling either the English or Continental rights, or $400,000 for selling both. If he sold the rights for more than these amounts he would receive 20% of the excess.

Hutchins sailed for Europe in mid-November. If Mergenthaler had read the 1889 report more closely he would surely have criticised the amount of commission offered to Hutchins.

## 10  *Reid's 1889 report and resignation*

For some time the Washington group had complained about the amounts assessed on their shares and the way that Mergenthaler had been treated. Reid was continually short of funds and stated in his 1889 report that on occasions they had to rely on private interest-free loans, all but two of which had been repaid. No doubt some of these were from his father-in-law, the banker Darius Ogden Mills who was on the Board. As the time for the 1889 annual report approached Reid was warned from several sources that the Washington group had formed a 'pool' that was trying to break the syndicate. This pool claimed that they controlled the majority of the stock, but said that they wanted to cooperate with the syndicate. Reid did not think they knew exactly what they did want – excepting dividends. [A reasonable demand because some investors had supported the inventors for over 20 years with no return.] Reid wrote to Rand that the Washington people were demonstrating, first, for the call on subscriptions to be halved, and second, for a change in the Executive Committee. Reid showed them that the assessment could not be cut and thought that they were convinced. It was reasonable and easy to change the Executive Committee.

Reid complained to Rand that some members of the syndicate were not taking their allocation of machines. In Rand's case he had taken pains to tell the Board that it was because they could not supply the required sort of machines, but Hine 'had made thrusts' at Smith and Rand about members of the syndicate who failed to take machines. He wished that somebody 'who wants an outfit would come forward and talk business', because 'we can fit out a newspaper with twenty machines on very short notice.'

When Hutchins had first contacted Reid in 1885 he estimated costs at about 5¢ per thousand ems exclusive of royalty and suggested placing machines in job shops, to avoid union problems. In December 1888 Reid, who supported the idea, wrote that four small pica machines were at work with plenty of cheap book copy to keep them busy. They were getting 25¢ a thousand for it, but Davids was sure that with new contracts they would get all they wanted at 40¢ a thousand. Reid was worried by the lack of skilled operators and wanted to develop the job office as a training centre. Mergenthaler ridiculed the idea because 40 poorly erected machines were crammed into a room only eight feet high, with gas pipes leaking, smoke pipes dripping, poisonous vapours escaping, 40 air blasts hissing, plus the heat of 40 Bunsen burners and the mechanical racket. 'A sawmill would be noiseless in comparison and a dry kiln not so hot!' Such a shop would attract only inferior compositors and 'lady typists'. [He admired Miss J. Julia Camp but was generally scathing about the competence of women.]

When Reid ordered Mergenthaler not to start work on a third set of 100 machines on 25 February 1888, it was not a cancellation, it was merely a pause for the company to review its position. The design of the Square Base Linotype was well advanced before the order for 200 Blowers was complete, and Greenleaf had written in such glowing terms about the new machine that the Executive Committee confirmed the decision.

Reid's report of 19 January 1889 stated, falsely, that Mergenthaler was in favour of placing large orders for the machine. In fact, as claimed in the biography, he had been appalled at the prospect of spending $260,000 on the experience of only 12 machines and pleaded for more conservatism, but was overruled. Reid mentioned the sale of the Baltimore assets for $20,000, claiming that it cleared the remaining expenses of the company except for $7,259.54 for tools belonging to Mergenthaler. Some tools were returned, but he had claimed that more than two-thirds were missing and while the committee was considering what to offer him he brought suit for the whole claim with interest and tried to attach what he assumed to be the company's property in Baltimore. He also claimed a 10% royalty on the cost of manufacturing the machines, unknown to some of the directors [although Mergenthaler raised this with Reid on 14 November 1887]. The manuscript of Reid's report contained similar criticisms of the Linotype to those in his 1888 report but they were deleted when it was marked up for the printing which apparently did not take place, so it is not certain that Reid really 'gave the machine a parting kick' as claimed by Mergenthaler. Reid noted that work on all parts beyond the 200 had been suspended, but stated that while the principle of the machines was accepted, the design should be modified. The new machine was not mentioned in the report, but that was a prudent policy because it was still in the design stage.

Mr McEwan, acting as secretary to the Board, called Directors of both old and new Boards to a meeting at 3.30 pm on Tuesday, 22 January 1889, to be held in Mr Reid's office on the ninth floor of the Tribune Building, and not at the shops in Brooklyn as originally intended. [The following paragraphs explain the change of venue.]

The stockholders were very disappointed with Davids's performance. Despite promising to produce machines cheaply he had assembled only about 40 between May 1888 and January 1889 costing $833 each in wages. Later calculations showed that each machine cost about $2,500. As usual, when Reid faced an unpleasant task, he delegated it; so, also on 22 January 1889, Lambert sent the following letter to Davids:

> At a meeting of the newly elected Board of Trustees, at which I was not present, held here this day, the following resolution was passed: –
>
> Resolved, That in view of the serious character of the charges that have been made against the management of Mr Charles H. Davids, the General Manager of Manufacturing, he be suspended until such charges can be investigated; and that in the meantime Mr Rice be placed in charge of the mechanical department, and Mr L. G. Hine invited to take charge of the business department.
> Resolved further, That the President of the Company be instructed to notify Mr Davids of this action by the Board.
> Yours,            E. Lambert, per D M
> PS The above was prepared for Mr Reid's signature but he did not return to town in time to reach the office today. As the trustees specially desired that notice of their action should reach you at once I take the liberty of forwarding it.

Mergenthaler claimed that the syndicate allowed the pool to take over so that Reid would not have to go to court over the *bona fide* sale of the company's property, but he was mistaken. Reid was re-elected but resigned gracefully because he was appointed United States Minister to Paris and could not manage the company effectively while serving the government. However, he continued to run the *New York Tribune* for many years as absentee editor.

He issued the following statement to the Board in February 1889:

> Gentlemen,
> Other engagements would, in any case, make it impossible for me to give proper time and attention to the duties of the Presidency to which you have done me the honor to re-elect me.
>
> But aside from this, it seems to me proper and desirable that those having the power should also have the full responsibility for the conduct of the Company's affairs.

I therefore respectfully tender herewith my resignation of the Presidency, to take effect immediately. In closing my service in this capacity, I venture to hope that the Directors, having now a job office measurably equipped, and in excellent position to begin lucrative work on book, pamphlet and legal printing, may not lose sight of the great opportunities this seems to offer. For three years we have been striving to reach and at last seem to have just about reached the point where we need only to operate our machines in order to take in three dollars for every one we spend in that branch of the work.

With best wishes for the success of the Company,

I am,     Very respectfully,     Whitelaw Reid.

Hine took over after Reid stepped down and Frederick J. Warburton, another member of the Washington group, was elected Treasurer of the Company. On 9 April 1889 he told Smith, who immediately wrote to Reid, that he had received a letter from Mergenthaler, stating that he had made a successful test of his new machine and was very enthusiastic about it.

## 11   *Opinion*

During this period the Mergenthaler Printing Company was not widely known outside the trade. Apart from a 40-page booklet produced in 1887, [the copy of which has vanished from the Library of Congress] there was no company publicity; some trade journals carried rumours of the Linotype, but there were no illustrations. This secrecy probably started when the syndicate voted themselves exclusive use of the Linotype.

Mergenthaler, who was undiplomatic, naïve and rather too quick to take offence where none was intended, accused Reid of debasing the Linotype, but this was not really fair. Reid reported honestly on the machine's shortcomings and its early unprofitability, but revised his opinion in the light of selling the manufacturing rights to the British. He wanted Mergenthaler to speed up production, reduce costs and be more compliant. At a time when the company was in financial straits he objected to the 10% royalty payment. However, until Mergenthaler's outburst in open meeting he was reasonably tolerant, but from then on, knowing that the inventor was short of cash, he used every boardroom trick to harass him by delaying payment for his tools.

# FOUR

## Hine takes over and Mergenthaler returns

The financial position of the company did not change when the new board took over. It was very short of cash, aggravated by syndicate members who had not taken up their full allocation of machines, but there was every prospect that Hutchins would solve the problem by selling the manufacturing rights to the British. Mergenthaler was pleased that the management had changed but was mistaken about the reason for Reid's resignation. At last he expected to be dealing with a sympathetic board. Greenleaf had told Reid that although the question of royalties remained unresolved, he could possibly persuade the inventor to drop the case against the company.

### 1  Mergenthaler reinstated

On 1 January 1889, Hine wrote to Mergenthaler that he would certainly have a better new year than the previous one; easy to believe because, for nearly a year, the inventor had received no income, had not been paid for his tools and machinery, had disposed of his shares to raise cash for his new factory and had been given the run around by Reid over the royalties to which he was entitled. Hine suggested that Mergenthaler should write two reports: one on the history of the Linotype from the start of the Bank Lane shop to the formation of the Mergenthaler Printing Company and a second covering the period from then until his resignation, stressing the conflicting orders and how his advice had been ignored. If Mergenthaler accepted this idea, Hine, as President of the National Typographic Company, would request such reports so as to absolve the inventor of appearing officious. It was noted on the side of the letter that the Washington pool would probably meet on Saturday, 6 January.

Hine wrote on 6 January that Mergenthaler's bill was *ordered* to be paid by Warburton, the new Treasurer, and that effective from 1 January last he was appointed Consulting Engineer at $4,000 per year, payable monthly. Hine also noted that Reid had resigned [not announced officially until February 1889] and ended – 'In haste for the 3.40 train.'

In some matters Hine was as cautious as Reid. On 6 March he wrote about being careful not to supply machines to people who could not afford to maintain them properly. He thought that he would find users for all the old model machines in the Brooklyn works and was anxious to get them out of the way as soon as possible. There was contention over the management of the factory. Davids had been ordered to hand over to Rice, who resented the subsequent appointment of Wilbur S. Scudder (see biographical notes) an instrument maker who had been employed by Mergenthaler in Baltimore.

On 9 March 1889 the *Scientific American* printed the first illustrated article on the Linotype, see figure 8. That popular science feature could have been a subtle attempt to advertise the machine prior to launching the British Linotype Company. It may also have been part of Hine's strategy to create demand in the USA because Reid had reserved all machines for the syndicate and had not invited publicity. At that time transatlantic liners regularly made the crossing in about ten days, leaving ample time for the publication to reach the UK before Hutchins opened his sales campaign.

According to Mergenthaler the Washington stockholders had taken up Reid's cry of 'too much royalty'. Soon after Hine took over, a committee of three of the inventor's most intimate friends (which may have included Greenleaf and Clephane) was appointed to see if he would accept a modified royalty agreement. Hine assured the inventor: 'If Mr Mergenthaler will consent to this modification he will so endear himself to the company and the stockholders at large that no request of his could be refused, and his position within the company will become impregnable for all time to come, the indirect advantages of which will be worth more to him in dollars and cents than the money he will give up under the proposed agreement.'

Mergenthaler yielded, and according to the biography, 'made the mistake of his life'. Later he commented that: 'Promises made by a company will last no longer than the term of office of those who made them.' On 25 March, Hine sent Devine details of the agreement reached with Mergenthaler. He wrote: 'In my conversation with Mr Mergenthaler I understood him to agree to the following as the basis of his relations with the company:

1  He will take the promissory note of the Company for $50 per machine for the 200 now completed and in course of construction, payable with interest in three years with provision that it shall be paid before that time if the company obtains the money by sale of patents abroad and as soon as money is so obtained.

2  He will take $50 per machine for those hereafter manufactured payable as soon as delivered to the user. If the new machine is an improvement on the present one he is to have a contract for making 200 or 300 depending upon the number he will undertake to make in a year, at a fair price. Of course we will have to rely on his judgement almost entirely as to price, payment to be made as the work progresses.

Not much was said about his personal services. The company will need them and, of course, must pay for them.'

Hine also sent a copy to Mergenthaler. He thought that the stockholders should approve the general features of the contract so that there could be 'no kicking in the future'. Hine wanted the position between the two companies and the syndicate to be resolved and believed that he was the only person with the time to be President of the Mergenthaler Printing Company.

Hine wrote to Mergenthaler on 24 April that Scudder seemed to be honestly trying to make a success of the factory. Nearly everything was ready, apart from pulleys. In Hine's opinion lots of 15 or more of the original machines could be improved without undue cost and made as fast as the improved machines. He thought that both models should be made 'by the thousand', but realised that his ideas were not generally supported. He expected to get support if the inventor would agree but would admit his 'mechanical stupidity' if Mergenthaler could prove him wrong. He wanted to convince Mergenthaler that it was 'our duty to get at least 500 machines at work earning money for the stockholders' before the next annual meeting on 15 January 1890. Hine's enthusiasm for both models of Linotype showed that he did not understand the technical problems. The new machine emphasised the shortcomings of the Blower, by overcoming them. If the company produced two incompatible models it would have lost any benefit from economies of scale.

On 16 May Hine became concerned about security. He had asked the inventor to show the new machine to Clarence Halstead, of the *Cincinnati Inquirer*, who had come to Baltimore to learn about the original machine, but thought that until the patents were out there was a great danger that pirates would delay and annoy them. While he did not want to restrict Mergenthaler's visitors it would be embarrassing if security made it necessary to exclude the Board and stockholders from the shops.

## 2   *The British Linotype Company and the American connection*

By mid-May 1889 Hutchins and his son Lee, an attorney-at-law, had set up an office and demonstration site in London. On 19 May, the *New York Tribune* printed an illustrated article in its Sunday supplement which gave details of composing room costs and claimed that the Linotype was saving the paper thousands of dollars a year, see figure 12. [The full text is given in Part II of the biography.] This piece was probably intended to support the campaign for launching the British company – there was no immediate need to tell *Tribune* readers about the Linotype. Mergenthaler was sure that Reid had suddenly decided to tell the truth about the machine.

During July Hutchins was granted legal authority to sell the overseas rights in Mergenthaler's printing inventions and was sent the assignment of rights for Jacob Bright, head of the British Linotype syndicate. Hutchins had hoped to conclude negotiations some two weeks after 13 July, when the British were due to start the press campaign to advertise the Linotype, but transactions actually took over a month. He cabled Hine: 'Matters settled fairly well. Sail on Saturday [17 August] by Umbria.' Hine presumed that Hutchins had sold the British rights for $2,500,000. Hutchins also sold 60 Blower Linotypes to the British for $1,000 each, claiming that as the actual cost of building the machines.

On 30 August, Hine informed Mergenthaler that Hutchins had sold the British patents for $2,500,000 of fully paid up stock in a $5,000,000 company formed in London. Therefore Hutchins's negotiations were disappointing in that they did not bring immediate cash to the Americans. Nearly all the stock was pooled so that any sales would benefit all holders, but Hine did not expect to see any money from that source for half a year. However, the sale of 60 Blower Linotypes would help the cash flow of the company in the short term by compensating for syndicate members who had not accepted their full allocation of machines.

Less than a year after its inception the British business was in a worse financial state than the American. The costs of setting up their factory had been much higher than the Americans had led them to believe and it was difficult to raise capital. Joseph Lawrence, as Deputy Chairman, went to

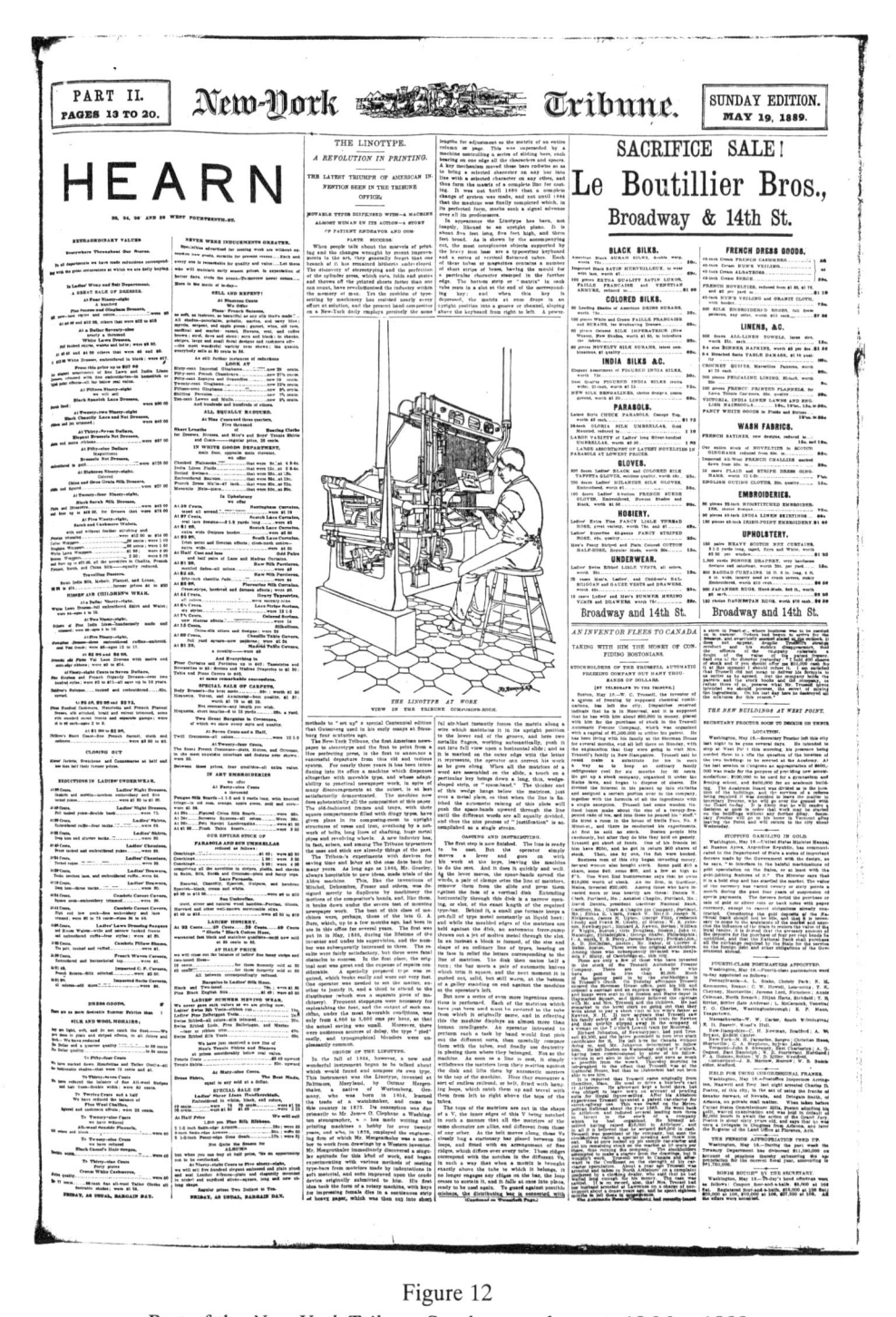

Figure 12
Part of the *New York Tribune* Sunday supplement, 19 May 1889

New York in June 1890, and negotiated a reduction of £450,000 in the ordinary share capital. It is not known how this was divided between the British and American interests. Lawrence saw the improved Linotype in July and cabled England to stop Blower production immediately.

On 27 August 1890, the British company authorised the cancellation of £450,000 ordinary shares and the creation of £100,000 of 6% Preference shares. Jacob Bright, who presided, said that an improved Linotype had been produced in America and claimed (without justification) that it would be possible to make Linotypes for £100 ($500) each.

There was friction between the British and American companies, possibly over the condition of Blower Linotypes sent from the USA. An agreement, of 21 December 1890, referred to a payment for outstanding accounts up to 31 December 1890. The Americans claimed $45,000 (£9,000) and the British counterclaimed for repairs. The Americans finally agreed to accept $35,000 (£7,000), in the form of 1,400 fully paid up 6% Preference shares.

On 27 January 1891, Hine asked Mergenthaler to send a machine urgently to The Linotype Company Limited in England to be delivered to their London office some days before 24 February for the stockholders' meeting. He was told not to worry about line length and not to hold things up over details. If he could make an extra mould of the length requested by Lee Hutchins then send it as soon as made. Three days later Hine wrote: 'The English machine is due to leave New York next Wednesday' [4 February]. On Tuesday, 3 February, Mergenthaler noted in his diary that he had sent machine number 328 to the English Linotype Company, London. Hine informed Mergenthaler on 10 March: 'Word from England is that the machine is running first rate. It started right off.'

Six months later, on 21 July 1891, Hine and the Board showed further confidence in the products from Mergenthaler's works when they ordered him to send one of his machines to Jacob Bright, Liverpool, England, by the White Star Line, no later than next Wednesday [29 July]. It was to be erected in the Manchester factory as a model for building machines. Hine knew that the inventor might have to take one of the machines intended for the *Commercial Gazette*, but left the decision to him. It was imperative to send a machine from Baltimore not later than Saturday night to be certain to get to the steamer from New York on Wednesday. They were going to charge $2,500 (£500) for the machine – rather more than Bright's estimate of £100.

## 3  General progress and production at Baltimore and Brooklyn

Mergenthaler made several personal visits to newspapers with Linotype installations. On 23 August 1889, Hine wrote to him after he had visited the *Louisville Journal* that he thought it very fortunate that the inventor had taken the rounds; no doubt that he would receive the same pleasant welcome in Chicago as he was given in Louisville. Mergenthaler objected to his likeness in the *Louisville Journal*, shown in figure 13, but Hine did not think it was very bad; it was probably as good as could be expected from a coarse cut on poor paper with a rotary press. In general, Mergenthaler's report was encouraging but Hine was disturbed to note that the machinist's bills on the *Louisville Courier* with 30 machines were over $80 per week while on the *New York Tribune* with 43 they were only $38, [possibly because the New York paper had more experience of Linotypes than the Louisville paper.] The *Providence Journal* machines were doing quite well. They had set 1,634,624 ems with 9 machines during 7 days of mid-August although not a third as much had been

Figure 13  Woodcut to which Mergenthaler took exception
*Source: Louisville Courier-Journal, 21 August 1889*

done on Monday as the average of the other 6 days. [An average rate greater than 3,300 ems per hour, assuming the 55-hour week current at that time.]

While Mergenthaler was busy visiting existing Blower installations and engaged on making the new machine less bulky, the Brooklyn factory was completing the remaining Blower Linotypes and starting to build the first 12 Square Base machines to Mergenthaler's design. As in the production of the original machine there was a requirement for special tools and problems due to parts that did not assemble properly.

On 13 March 1890 Mergenthaler proposed building machines for the company. Hine replied on 17 March that he was sure that he could get the majority of stockholders to support the proposal. He thought that although money was short they could get plenty with present prospects. He did not expect the shareholders to 'growl' much at further assessments and they might not be needed. In April, Mergenthaler was awarded a contract for 100 machines at $1,200 each, matrices and space bands extra. A similar order was placed with the Brooklyn works.

In the meantime the Brooklyn plant was finding it difficult to complete the first dozen improved Linotypes. The inventor made several trips to help and finally sent two of his best men to work in Brooklyn. The first six machines were delivered in October 1890 to the composing room of the *Brooklyn Standard Union,* the organ of the Typographical Union. This first commercial installation of improved Linotypes was vital for the success of the company because its adoption of the Linotype ensured union recognition for the machine throughout the United States. Hine said that they would be running with fairly good operators on 6 or 7 October. Of course there were teething troubles. Hine reported on 11 November that only four of the *Standard Union* machines were working regularly; the other two were in the hands of the assemblers. On 21 November he expected to give the *Standard Union* control of their machines; they were now all running and doing better almost daily. Mergenthaler blamed these initial problems on defective workmanship and inexperience at Brooklyn but praised the early operators who had all been recruited from the regular staff. Every one of them soon made a first class record for himself and most went on to become record breaking Linotype operators.

By 24 November the Mergenthaler Printing Company put an illustration of the Square Base Linotype on its stationery, to show that it had become the standard model, see figure 14. In his letter of that date Hine apologised for sounding gloomy he 'was never personally in better spirits'. He continued:

'I regret to hear you are not in the best of health. It will not pay to use that up, and worry will do it faster than work. You have too much sense to worry, or become disconcerted.'

By the beginning of 1891 Hine was becoming very concerned about output from the Brooklyn factory and wrote to the inventor:

> Dear Mr Mergenthaler, Another year has passed and not a machine built. Whenever any work goes from the Brooklyn Factory to you that is not well and correctly done please send it back and advise me wherein it is wrong. I hope soon to learn how it happens that we cannot build a machine so it will work – without spending as much time and money in tinkering it as it ought to cost to build it. There is nothing specially creditable to report from any of our operations here.

Mergenthaler had intended to start deliveries under his new contract in October 1890, but schedules slipped due to the problem of tooling up for production. This raised costs which that meant that he would lose out on the contract, but as the inventor he was able to arrange matters to minimise the loss, while the machines from Brooklyn cost over $2,000 each. On 19 January 1891 he wrote in his diary: 'First of the 100 machines completed and working to entire satisfaction.' The Brooklyn factory did not deliver any machines until February 1891.

Mergenthaler claimed that the new machine immediately became much more popular than the original model. Although the yearly rental was $500, as against $300 for the Blower, everyone chose the later machine. The inventor complained that this faster and more popular machine was depriving him of royalty income to which he would have been entitled under the original royalty agreement.

Some measure of Hine's confidence in Mergenthaler is shown by his letter of 27 January 1891, only eight days after the first machine was finished, which gave the inventor just a week's notice to send one to England. [As noted in section 2, he met the deadline and the machine worked perfectly.] However, Hine was disturbed about output from Brooklyn; a machine was expected to go to Canada the next Saturday, 31 January. Hine commented: 'If not done it will be fair evidence there is a *screw loose* here, the delay is becoming intolerable.'

There was evidence from the biography and letters to Mergenthaler, that Hine was becoming increasingly frustrated with repeated delays and tried, like Reid before him, to establish regular working practices in the factories. On

30 January, Hine wrote that things were not going well and that they should try to learn from past mistakes. He asked Mergenthaler to send working drawings of everything necessary to build machines exactly as he built them; if they were not followed exactly and thereby blundered to the prejudice of the company it would be their fault and no blame could be attached to the inventor. He was pleased that Mergenthaler had found the right man to install the machines made in Baltimore and assured the inventor that he was the right man in the right place. He was therefore, with the approval of the board, passing the three largest southern orders to Mergenthaler. These were the *New Orleans Times-Democrat,* the *Memphis Avalanche-Appeal* and the *Cincinnati Commercial Gazette* (which Hine described as the most notorious paper in the West).

On 9 February 1891 Hine ordered Scudder to complete the Linotypes now under construction strictly like the standard machine just tested and to make no changes without written authority. A similar order was sent to the inventor and copied to the stockholders with the explanation that it was necessary because production at Brooklyn had been delayed by the many changes made by

Figure 14   Letter headings used by the Mergenthaler Printing Company
*Source: Photocopies from letters among Mergenthaler family papers*

Mergenthaler, who obviously thought this to be unfair comment. He had the best interests of the company at heart and 'was always anxious to get the machine into the highest state of perfection and also reduce its cost wherever possible.'

Mergenthaler believed that he was the only rational member of the enterprise and showed his irresponsibility by ignoring the order. He claimed that the final machines produced in his works were neater and smoother than those made in Brooklyn, and that his improvements and economies had reduced the average cost of each machine built by him to little over $1,100 whereas the Brooklyn machines were jerky and noisy, at least six months behind the state of the art and cost over $2,000 each. He declared that the company was ready to adopt his improvements 'for use in the next lot of one hundred machines and so it has kept on to this very date.' He reasoned that Hine, as a lawyer, would expect a trainee lawyer to use his experience of one court case to argue another, but did not realise that he was not comparing like with like. Similar court cases will differ in detail – a mechanic who maintains machinery needs uniformity of design to work effectively.

On 15 March, Hine wrote that it was absolutely essential not only to produce machines but to sell some. He asked Mergenthaler when the first ten machines would be out and then how many per month; this forecast of output would affect future policy. Brooklyn expected to produce 15 to 20 per month and the inventor apparently promised 3 or 4 machines per week.

Hine tried desperately to cajole Mergenthaler into seeing reason. In a 2-page letter dated 29 March he wrote: 'If the index to working drawings on three sheets appended to my report to the board of directors of March 13th 1891 is not correct, the errors should be pointed out. If it is correct then not only the board should have it but the stockholders are entitled to it as a matter of absolute right. It is too late for anyone to find fault with you. No one but a knave or a fool will suggest that you have not devoted yourself with wonderful ability and industry to solving the problems of type composition by machinery commenced by you *when a boy*, [poetic licence?] and if we had known as much about it six years ago as we know now at least $600,000 could have been saved, yet the thing has been accomplished and it is worth at least what it has cost. There is no way but to close the books to the past and show our faith in our own judgement that the machine is fit to build by building it as it is. That improvements will come is not questioned but they should be first tested on an independent machine before adopted.' Hine asked about the state of orders for the southern papers and suggested that Mergenthaler should personally set up the machines in New Orleans. He concluded with optimism for Mergenthaler and doubts about his own future: 'It will not be long before the company will be able, as well as disposed, to fully satisfy you for personal services so far as money is concerned. I hope to see to this before I am turned out, or find my efforts so criticised as to make my position unbearable.'

There was a gap in the correspondence until Hine wrote on 8 May: 'It has been so long since we heard of your progress that it is getting lonesome.' Hine noted that there had been no word from New Orleans and that two machines were to go the *Staats Zeitung* next Tuesday [12 May] – if the latter machines on test by a German worked well.

Hine sent Mergenthaler a sample of German setting from the *Staats Zeitung* on 13 May 1891, and noted: 'Two machines will go in the *Staats Zeitung* tomorrow if they (the paper) is ready. We are ready.' In that case the customer kept the company waiting. Hine had also seen Richard Smith of the *Cincinnati Commercial Gazette*, who wanted ten nonpareil and five minion machines. Hine asked the inventor, 'Please make them that way.' However, as all Linotypes after the Blower could carry a range of type sizes it must be assumed that Smith was referring to the fonts of matrices and the space bands to be supplied with the machines.

On 22 July Hine wrote that the Board had met and discussed how to raise enough money to build machines in large numbers. They wanted to produce 1,000 machines as soon as possible by contracts and had adjourned until that morning but as a quorum had not appeared by noon they adjourned until the following Tuesday [28 July]. Mergenthaler's proposition to build machines was laid over until then – as was everything else of difficult account! A minority of the Board wanted to break up the Brooklyn factory which would have been a serious mistake.

Hine continued to worry about security, particularly about plans for future models that might impact current sales and the possibility of being pre-empted on the invention of improvements. He raised two confidential technical points on 28 July and 6 August. First, they should apply for a patent on an automatic feeder of metal into the casting pot. This had been discussed at least two years earlier and if they did not act now some one would have a patent on it and offer to sell it to them. Second, he asked the inventor to go ahead with his improvement of the keyboard and prepare one machine in the quietest way possible. Hine explained: 'We cannot afford now to have it understood that we are building an improved machine on the present improved machine. We ought, of course, to make the best machine possible but not, in the slightest degree, interfere with, or delay, the building of the Linotype as it is now.' However, Hine did not think that they should experiment with steel matrices at that time.

Problems arose in New Orleans in the summer of 1891. Hine informed Mergenthaler on 27 August: 'I wrote to *Times-Democrat* about our bill. They replied pleasantly and said they gave Mr Letsch a pass to Cincinnati and that the charge should be cut. The *Times-Democrat* also sent a record of the machines showing three of them in bad condition but attributed it to the incompetence of their machinist and asked us to send a thoroughly competent one to them to remain permanently. Can you send them such a man? It would be an excellent thing to do.'

By 8 September the position was so desperate that A. G. Winterhalder, business manager of the *Times-Democrat*, sent Hine a typewritten letter saying that they wanted a first-class man who was competent in every respect and they wanted him badly. Their machinist had about given up; he was a good mechanic, but the linotypes were too much for him. They were going from bad to worse; for the past three or four nights from five to seven of them were knocked out entirely and they had to use hand composition to get the paper up. The working of the machines for the past two nights was worse than during the second week after they were installed. They needed a man who could put them in order with the least possible delay. 'As they are working now, they are practically valueless.'

Hine appended the following hand-written letter to Mergenthaler to the typescript from New Orleans on 12 September: 'The above is a copy of a letter just received from the *Times-Democrat*. It is of course of the first importance to make the machines now out run well and satisfactorily. We cannot exist without doing this and if those who take the machines cannot, or do not, keep them in condition to do steadily the work they are made for there is no alternative but to furnish a man who can and will do so. [At that time many people did not realise that mechanical systems need regular maintenance.] The *Times-Democrat* is very liberal in wages, offering $30 a week for a *competent* man. We send machines out in such small lots that all our men of any experience are out. It would, too, probably be uncertain in results, if we sent a man to tinker with machines built by you.'

That letter showed that Hine realised that only Mergenthaler or a man from his shop would be able to fix the machines at the *Times-Democrat*. There is no record of when, how or whether these problems were resolved, but Hine's later letters showed that most installations settled down more successfully than the *Times-Democrat*.

When Hine wrote to Mergenthaler on 5 November he was thoroughly fed up with being blamed for other people's shortcomings. Machines that were scheduled for the *Denver Times* would have to be held back. The contract was sent to that paper through Wm Henry Smith, who placed the machines there, and Hine understood that they were wanted as soon as possible. The paper returned the contract and changed the delivery date from 1 November to 1 December, which was of course, as Hine wryly remarked, further evidence of mismanagement on his part. He was glad to know that the opportunity for making such complaints would soon end.

## 4  *The Typograph and the start of litigation*

In the biography Mergenthaler claimed that so far the invention had been free from imitators and infringements, but 'it is a bad thing which does not find infringers.' In 1890 the Linotype was challenged by a new linecaster, the Rogers Typograph, described by Mergenthaler as a bold and barefaced imitation of the Linotype and a comparatively ineffective machine. Rogers always claimed that he was unaware of the Linotype when he invented the Typograph. He may have been following current trends when he decided to make a slug casting machine which his backers probably thought had sufficient novelty to be granted a patent.

In general appearance the Typograph, as shown in figure 15, resembled a large typewriter and, like the Linotype, composed lines of individual matrices. These were threaded on to a system of wires, one for each character, and did not circulate freely. Typograph spaces were circular wedges which were rotated to justify the line. Matrices released by operating a keyboard were assembled by gravity, a distinct advantage over the air blast of the Blower Linotype. The operation of the machine resembled that of the second band machine rather than the Blower because lines were assembled and cast at the same point and the operator was restricted to the characters on the wires. After casting a line the operator upended the system of wires to distribute the matrices by gravity. The makers claimed that this interruption which potentially made the Typograph much slower than the Linotype was really an advantage, because the operator could use the three-second delay while the line was being cast to read the next

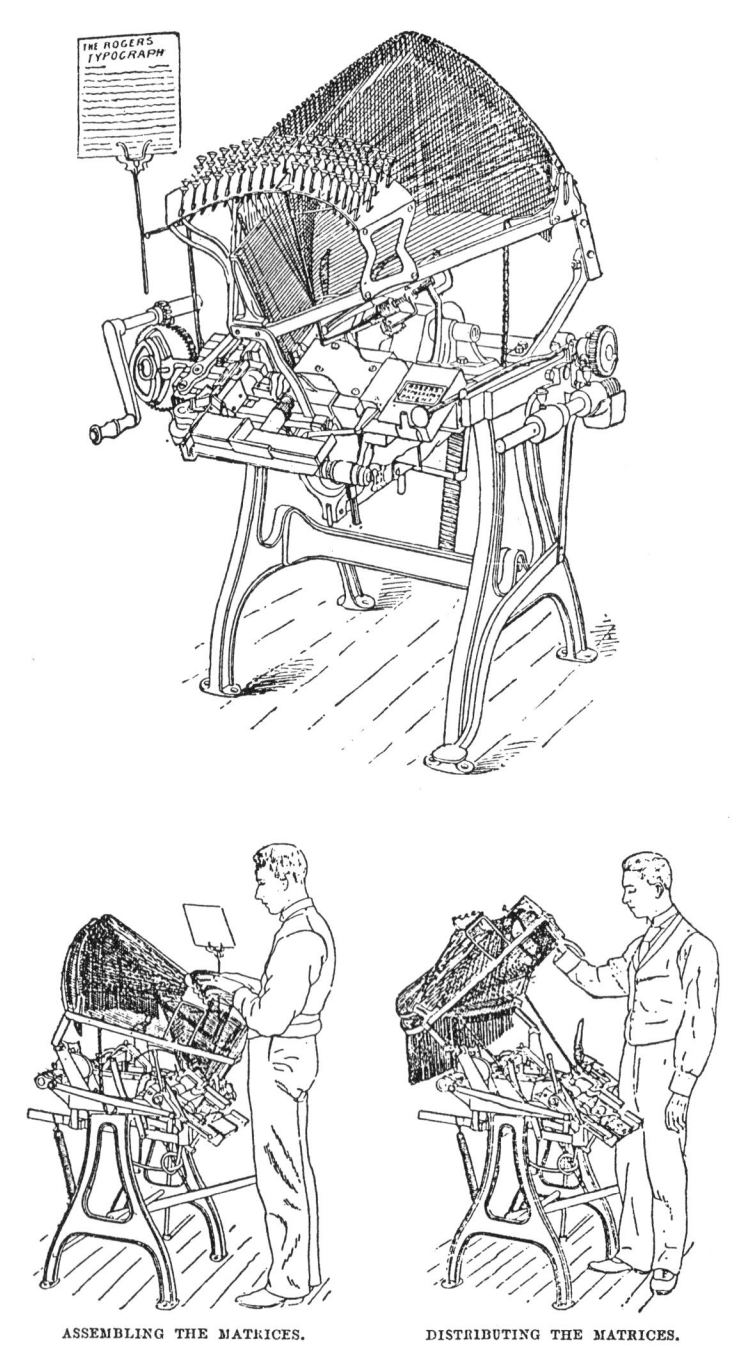

ASSEMBLING THE MATRICES.   DISTRIBUTING THE MATRICES.

Figure 15   The Rogers Typograph
*Source: British and Colonial Printer and Stationer, 1 January 1891, p 2*

line of copy! The Typograph operator stood in front of the machine, like a hand compositor, which may have been convenient for distribution, but was quite unsuitable for operating a keyboard effectively. The operators of all other mechanical typesetters sat at the keyboard. The machine required one-eighth of a horse power to be driven mechanically but could also be driven by hand or with a foot treadle.

When the Mergenthaler Printing Company first heard about the Rogers machine Dodge went to Cleveland, where the Typograph was produced. He said that the machine was a 'palpable imitation of the leading features' of the Linotype and threatened legal action against anyone who attempted to sell or publicly operate any machine that infringed their rights. The first American Linotype advertisement, as shown in figure 16, appeared in the August 1890 edition of the *Inland Printer*. It threatened the users of the Typograph and other competitive linecasters rather than the producers of the equipment and only mentioned the performance of the Linotype as an afterthought. This advertisement appeared more than a year after the Linotype had been widely advertised in the United Kingdom.

These notices did not stop Rogers and his backers from introducing the machine. In September 1890 it was exhibited at Pulitzer's newspaper, the *World*, which became the sales office of the Typograph Company. It was

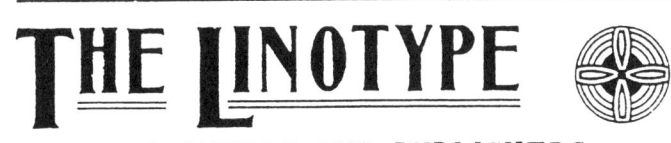

# THE LINOTYPE

## TO PRINTERS AND PUBLISHERS.

All known Linotype Machines, and the product therefrom, are covered by Letters Patent Nos. 362,987, 313,224, 317,828, 345,525 and other patents controlled by the undersigned company.

The public is cautioned that the use of any machine which casts, as a substitute for movable type, linotypes or type bars, each bearing the characters to print an entire line, unless purchased from this company, will render the user liable to a suit for infringement.

The Linotype Machine, made by this company under its patents, is now for lease or sale; is capable of an average speed of 8,000 ems per hour, and the print from its product is superior to that from movable type. Any size of type from agate to pica can be produced upon the same machine. We earnestly invite your investigation.

For full information address, or visit personally,

**THE MERGENTHALER PRINTING CO.,**
154 Nassau Street,     -     NEW YORK CITY.

Figure 16   The first American Linotype advertisement
*Source: Inland Printer, August 1890, p 1039*

reported that the company had received orders for 600 machines, a claim that cannot be verified. It was also claimed that in setting up eight pages of a copy of the *Sunday World* the machine was run for 125 hours without any interruption. The start of the Rogers sales campaign coincided with the Mergenthaler Company abandoning its exclusivity arrangements and starting to deliver the first Square Base Linotypes to papers outside the original syndicate.

On 5 October Hine wrote to Mergenthaler that he had seen the machine two days earlier. It worked well and easily, was a trifle more than half the speed of the Linotype and cost one quarter as much to build. There was no danger of reaching a compromise with Rogers. Hine expected to file a Bill of Complaint against the Typograph early in the following week. 'That they infringe is hardly open to question.' In fact the papers were not ready for five weeks. On 11 November he wrote that Dodge had been preparing to proceed against Rogers's 'rattle-trap' and thought that the Bill would be filed against that concern the next day. He continued: 'We will squelch them in a few months and I believe will so expose them at once that they will be powerless to do much mischief to the public. They have done no damage to us further than to discourage our stockholders and distract attention from our regular business.'

The obvious potential customers for the Typograph were newspapers that were not supplied with Linotypes because they were in the same town as a syndicate member. It was reported that the Typograph in America had a keyboard like a Remington typewriter, but the British keyboard was not in QWERTY form. In view of its limitations the Typograph was quite expensive by comparison with the Mergenthaler machine; it was lighter (450 pounds against a ton for the Linotype) and the mechanism was simpler. Its price of $2,500 was little less than the $3,000 for the Linotype, despite costing only a quarter as much to make. It took 20 minutes to change type size and line length.

The Mergenthaler Company submitted affidavits late in 1890 that there were about 170 Linotypes in use with a further 200 under construction. Mergenthaler stated that he had been developing his machine for fourteen years, that there was no other similar machine of its type in use, or before the public when he introduced it. He had inspected the Rogers machine in New York and considered it to be an imitation of his own, that it would be an invasion of his patents and a threat to his business if permitted to enter the market. Hine wrote to Mergenthaler on 30 January 1891: 'We are having a warm time with the Rogers Typograph. Two experts and others not professional experts swear to no infringements, etc. We are busy now showing they are asses – or we are.'

On 11 March, Mergenthaler noted in his diary that a preliminary injunction had been granted against the Rogers machine. Judge Lacombe ruled that even if the Rogers machine was 'lighter, smaller, cheaper, more easily operated, and more efficient' that was immaterial if the Linotype was covered by a foundation patent. The injunction stopped exploitation of the Typograph for the time being. In May the Rogers company shipped its machines back to Cleveland from the Pulitzer Building, appealed against the ruling and established factories in Canada and Germany where the injunction did not apply.

Hine wrote to Mergenthaler on 13 August: 'The taking of testimony by the Rogers people will soon get hot and probably you will find persons falling in your way to "pump" you as to matters connected with your inventions.' He continued: 'Of course you need not be reminded that nothing you can say outside of examination in court, or to counsel for the [Rogers] company can be of any use in the litigation, but anything you should say (or be understood to say) that could be tortured against you could be used to your prejudice. So please do not talk to any one but counsel about your patents, or attempt to give dates, or answer questions.' Hine wanted Mergenthaler to come to New York in September to have a full talk with Mr Betts [of Betts, Atterbury, Hyde and Betts of New York]. He ended: 'I have no idea you would talk about the case or your patents to other than counsel but simply take the liberty to write a word of caution.'

On 27 August Hine wrote that Mr Betts wanted to take Mergenthaler's testimony on 3 September and that it was important that Mergenthaler meet him then. In his diary entry for 3 September Mergenthaler confirmed that he had had his first conversation with Mr Betts. He was in New York again on 8 September to testify in the Rogers case but it was postponed on a point of procedure; therefore he spent most of his time at the *Staats Zeitung*, *Commercial Bulletin* and *Morning Journal* where the machines were 'generally unsatisfactory on account of keyboard and lack of cleanliness!' [As a watchmaker he would have expected the mechanics at the newspapers to have maintained their machines properly.]

The Typograph won a six-day trial of type composing machines run by the American Newspaper Publishers Association in November 1891. The machine had been set up in 90 minutes and, under the sole control of the operator, had run smoothly for five consecutive days. The Linotype did show 'bursts of speed exceeding the capacity of its competitors', but lost partly due to reckless operation. The third machine was the McMillan, a machine for setting movable type; it was typographically excellent, but the need for three skilled workmen made it unsuitable for newspaper offices.

Another problem arose at the end of Hine's term of office. Schuckers and Mergenthaler had both applied for a patent on a double wedge spacer early in 1885. The Patent Office delayed issuing a patent until late in 1891 when the commissioner for patents ruled in Schuckers's favour. Hine said the decision would have no immediate effect because the company would be appealing it. He also claimed that the double wedge was not necessary because the company owned other equally practical devices.

## 5   *Financial considerations during Hine's term of office*

Hine was continually short of cash during his term of office and the stockholders were becoming disillusioned with a Board that made regular assessments but paid no dividends. Throughout this period Hine's letters to Mergenthaler were friendly and encouraging and he made regular visits to the inventor. By the end of 1889 when Hine was becoming very concerned about finances the inventor sent him a cheque for $187.50 in connection with stock and on 27 December Hine replied: 'I have not been so nearly dried out financially for years but am not hard up enough to have urged you for payment under present circumstances.' Little or no cash was coming from the United Kingdom, the 60 Blowers supplied to the British were largely paid for with preferred stock and the American companies had taken a cut in the amount of ordinary stock in the Linotype Company after Lawrence's New York trip in June 1890. Production was delayed and extra costs were involved to equip the Brooklyn works for manufacturing the improved Linotype. The few machines of the new model that were made suffered from teething troubles. Letter after letter from Hine to Mergenthaler told of the large outflow of cash and the need for economy.

On 21 November 1890 Hine told Mergenthaler that it was doubtful if they could do more than pay necessary current expenses during the next two months. Stockholders were not responding promptly enough to calls and it would probably be impossible to get help from the bank during the present financial flurry. So he asked the inventor to be as economical as possible, consistent with maintaining schedules.

On 3 December, Hine sent a copy of a letter that Mergenthaler had written to him on 24 January 1887, with the message: 'This letter perhaps will carry you back to your search 4 years ago and aid somewhat in fixing dates.' Hine also sent a copy to Dodge but left out a sentence in which Mergenthaler suggested that claims on the wedge spacer should have been in the early patent application. [This seems to have been just one of the cases where Mergenthaler claimed an oversight in Dodge's work as his patent attorney.]

On 8 January 1891 Hine wrote: 'I quite agree with you that the situation is not gloomy – it is simply interesting and unquestionably requires careful provision for future contingencies.' Costs were more than $1,000 a day and there was seldom cash in hand to cover the week's expenses. There was practically no income excepting that drawn from stockholders, most of whom had to borrow money to pay calls, all must look out for a squall and prepare for it. [This was less than tactful since the inventor had to sell his holdings to establish his new factory. Hine's 1885 prediction that the Mergenthaler Printing Company would not call for more than a 25% assessment had been completely wrong.] Hine wrote: 'When we cannot pay our bills for work and material we must stop' and then tried to soften the statement by saying that it was an unlikely event, but claimed that the inventor was entitled to know the true position. Mergenthaler should not be troubled about money matters which were the business of the Board and there would be plenty of money if they could get machines. Using the excuse that he wanted to report fully to the stockholders on 17 January he asked for a detailed progress report and predicted delivery dates. He was just letting Mergenthaler know that they were going through a bad patch.

By 27 January they were so pressed for money that they could not pay royalty on the new machines. They were not receiving a dollar for them and most were running at the company's expense. Hine told Mergenthaler that it would soon change but, with expenses considerably more than $1,000 per day, there was not enough on hand to meet the next pay roll. Whatever may happen, Mergenthaler would of course be paid for the contract work as agreed, but they could not pay royalty at present. 'Sorry to say so but the truth must come out.'

Hine was getting to the end of his tether on 15 March when he wrote: 'It is now absolutely essential that we not only get out machines but sell some. We are now on our last 3% call and will have to concentrate our efforts on a few machines in each factory and get them out rapidly for a time.' He was even more blunt on 21 March 1891 when he told Mergenthaler that the general feeling in the board was that he ought to have a salary commensurate with the revenue of the company when it derived any revenue but while the inventor was doing contract work and the company was expending instead of receiving money it could not in fairness to stockholders pay more than the royalties on the machines that would earn something. He ended that they must pay their weekly bills, of course, or quit. So, once again, when the financial pressure was on, Mergenthaler was the one who had to wait for part of his payment.

Hine was optimistic about finances at the end of March through to mid-May but they remained on a knife edge for the next few months. On 22 July he said that there was not enough money in the treasury to pay the next week's probable bills. The Board had passed a resolution to advertise all unpaid stock for forfeiture, on 1 August. He wanted to know if the inventor could manage on $1,500 that week although the Board had agreed to pay $2,000 weekly for two weeks. He ended: 'Those who insist on drawing conclusions that the management has been inefficient and extravagant from the fact that a large amount of money has been expended and few machines produced, without investigating the causes, will soon have an opportunity to prove their capacity for doing better.'

The reason for Hine's confidence in the future became apparent on 6 August 1891, when he told Mergenthaler about plans to reorganise the company as the Mergenthaler Linotype Company of New Jersey, to take in the United States and Canada with a capital of $5,000,000 of which $1,000,000 common stock would remain in the treasury for future use. A general stockholders meeting was scheduled for 14 August. Mergenthaler criticised the plan to consolidate the companies as watering the stock to the extent of three additional shares for every two existing before. He claimed that one million of the new stock was soon sold to a syndicate, including D. O. Mills and Ogden Mills (respectively father-in-law and brother-in-law of Whitelaw Reid) for about one-third of its face value. [It is not known whether this was the full amount paid or just a preliminary assessment.]

On 19 August Hine wrote: 'We did not have a quorum of stockholders last Saturday.' [That was 15 August, but Hine had said that the meeting was scheduled for 14 August. Could the shareholders have been misled about the date?] Yet again there was too little money in the treasury to meet next week's payroll, but of course they would have to raise it. The letter ended: 'It is believed we will be able to float our new stock when issued. The company is organised but what complications will spring up before it is got to going is yet uncertain.'

On 21 August Hine reported that the stockholders had complained about assessments while prominent members of the company had not paid their calls. He wrote that they were very close pushed. A number of stockholders said that they would not pay until Hutchins and Bryan [an original member of the

Washington group coopted on the Board by Reid] had paid their outstanding calls of about $26,000. He said that he would be advertising all delinquent stock very soon after 1 September.

Mergenthaler noted in his diary on 1 September, that Hine had called him to New York on important business. The following day he made the acquaintance of Mr Mills (the old gentleman) and was present at two board meetings with Messrs Hine, Bryan, Clephane, Warburton, Dunbar and Mills junior where they discussed raising money to build further machines. Mergenthaler told them that he would have to insist that his royalties were protected.

Hine wrote to Mergenthaler on 28 October: 'We have a meeting of the Board tomorrow and it is expected then to settle practically the scheme of reorganisation and my relief from denunciation by parties who assume to know more by intuition and inspiration than all the active workers learn by unremitting attention to the work of building the machines and their introduction. The work of the past year nets the company at least 30% on the amount it cost. There has never been a dollar received for annual rent of machines excepting those put out while I was general manager – two and a half months in the spring of 1889 and since Oct 1st 1890. This may be regarded as an accident, or a coincidence, but it is sufficient reason why I shall decline to reply to charges of mismanagement – at least until some one else shall have done better – a comparatively easy thing to do now that the preliminary work has been done. The probability of a change in management compels me to decline to attempt to [make] any contracts, as suggested in yours of the 26th, or to give directions at present for future work. I will write you more fully next Friday. The 'Chicago Trat' and why we failed I will publish to the stockholders as soon as I get the facts. The machine left the factory as good a machine as was ever turned out.'

On 31 October 1891 Hine informed Mergenthaler that the meeting of the two boards last Thursday [29 October] had agreed the general plan for merging the two companies into the Mergenthaler Linotype Company. Five days later Hine wrote that D. O. Mills had just told him to make the contract with Mergenthaler as proposed and said that if the inventor had the form of contract drawn, to send it on; if not, Hine would draw one and send it to him – or better still they could sit down together and draft one in a personal interview.

The inventor claimed that control now passed from the Washington group into virtually the same hands it had been in before Hine took over from Reid and that from then on he was subjected to the same ill will he had previously experienced.

In the last of these letters to Mergenthaler on 16 November 1891 Hine wrote: 'The conditions now are so very different from those of a month ago that I think it better for you, and more prudent for me, to wait until the new and *improved* order of things takes place before deciding on the construction of the machine as you have lately improved it.' He had sent copies of Mergenthaler's letters to Ogden Mills and to D. O. Mills, who had since left for California. Hine was sure that they would choose the very latest improvements. This letter ended with the caustic comment: 'Will have command of my own time on and after the 25th inst and don't believe I will ever again work for wages with a board of directors or trustees for boss. It may be because I will not get the opportunity!'

Hine left the company in December 1891 and Dodge was elected to replace him as President and General Manager.

## 6  Comment

The cordial relations between Mergenthaler and Hine between the years 1889 and 1891 exemplify the old adage that 'you don't catch flies with vinegar.' Where Reid ordered, Hine asked. Hine often kept the inventor waiting for payment when funds were short but, unlike Reid, always kept him in the picture and did not delegate his communications to underlings. Mergenthaler appreciated this consideration and accepted suggestions from Hine where he would have challenged directives from Reid. The rapport between Hine and Mergenthaler probably helped to establish the Linotype enterprise when it might have failed from technical and financial problems under different management.

## FIVE

## *Dodge's rule and contention from other linecasters*

The Linotype had reached its definitive configuration when Dodge took over from Hine at the end of 1891. Future developments would focus on improving the printing surface and the flexibility of the machine. Dodge had an easier task than Hine and Reid because two fully equipped factories were producing increasingly reliable Linotypes. There were 160 machines of the new model already in service, 52 ready for delivery, and 100 nearly complete at the Brooklyn works. There were $374,000 in the treasury and an annual rental revenue of $80,000. Clephane kept Mergenthaler posted about events in New York after Hine left the company.

At last production was on an even keel but it was not the end of the company's troubles. When Schuckers was granted a patent on the double wedge spacer it put the whole Linotype enterprise in jeopardy and gave the Rogers Typograph Company a means of attacking the Mergenthaler Linotype Company. Scudder, Superintendent of the Brooklyn factory, left to exploit the Monoline, his new linecaster.

### 1  *Developments in the Typograph case*

The Linotype company continued its action against the Typograph, but the Typograph company defended itself by acquiring a potentially lethal weapon. In mid-1892, two months after the Schuckers justifier was awarded a patent, the Rogers Typograph Company gained control of that patent by merging with the Electric Typographic Company of New York, NY, who owned the patent, to form the Rogers Typographic Company. [In September 1885 Hine had been offered a major share in the device for $50,000 but it was rejected, largely on

Dodge's advice.] The owners of the Schuckers patent began to hint at suits for infringement over the use of the space band – this forced the inventor and the Mergenthaler Linotype Company to develop the devices about which Hine had boasted. Following those innuendoes the editor of the *Journalist* wrote that the Mergenthaler people would 'now have an opportunity to . . . taste the sweets of injunction.' In July he reported that the Mergenthaler people claimed to 'have their pockets full of justifying devices which are better'; but those who had seen them had asserted that the 'justifying spaces have a confusing habit of getting soldered together in casting the slugs so that they have to be distributed with a hammer. This mitigates against rapid composition.'

Dodge was very pessimistic and on 11 July 1893 Mergenthaler wrote in his diary: 'In regard to the Schuckers [patent] Dodge expresses himself as having no hope of carrying the case successfully through the Court and much to my disgust says that we would be compelled to shut down the works and also be forced into consolidation with Schuckers on the basis of equal partners. Told him that the proposition was outrageous which he denied, saying that if we needed their space band it was worth as much as the whole machine otherwise. His position that we absolutely needed it does not seem to be justified by condition of substitutes now at the command of the Co and I told him so.' Mergenthaler continued with his experiments and produced at least three versions of his wedge spacer which, judging from uncorrected proofs pasted in his diary in September 1893 seemed to work well.

Dodge continued to press Mergenthaler for a practical substitute for the double wedge space band and in a letter dated 13 July 1893 threatened to withhold the inventor's royalties on the grounds that without the space band they were not getting a complete machine. Mergenthaler commented in the biography: 'Another one of Mr Reid's contributions to the castle on the Rhine,' and replied that his relations with the company did not depend on the 'opinions and sentiments of its officers,' but were regulated by a written contract which he intended to enforce. The company carried out its threat; it ignored every demand for payment of royalty for eleven months.

In mid-1893 the injunction against the original Typograph (not to be confused with a later Typograph system consisting of three machines) was made permanent. Judge Alfred Coxe gave his opinion that no previous composing machine anticipated the Linotype, that Mergenthaler's patents should be liberally and broadly construed and hold as infringers 'all who produce a linotype [slug] by similar or equivalent combinations' and that the inventor had made an invention of unusual merit and was entitled to reap the reward. Litigation over the Typograph finally ended in 1894 when Judge Marcus Acheson cited the Coxe decision and affirmed a New Jersey district court decision in favour of the Mergenthaler company.

On 31 October 1893 Mergenthaler wrote in his diary: 'Dodge sees the justification No 3 is pleased with result and working of it. There is some talk about building 25 machines with step justification but no final decision about it. Dodge like usual says, "It ought to be done".' There is no evidence of Dodge placing an order for those machines.

By January 1894, Mergenthaler started to produce his step justification (described in section 2) and built 225 machines fitted with this modified spacer at Baltimore. He claimed that, despite problems, the new justifier was effective and should have been used as a lever to bargain with the owners of the Schuckers patent.

While Mergenthaler was working on his justifier in Baltimore Dodge was developing another system in Brooklyn. However, a machinist in Brooklyn thought that it was not strong enough. Clephane wrote to the inventor on 23 February: 'I do hope your justification will work all right. Dodge, in response to an inquiry at the last meeting, stated that he was building another machine with his justification feature in it and he expected first class results. I said, "Mr D, Mr M is also building machines with his new justification feature in it, is he not." "Yes," he said. I was determined that the Board should know that you were working on it, as well as he, and I have no doubt you will beat him.'

Clephane tried to get Dodge to give Mergenthaler a contract for at least 100 machines. After seeing Dodge on 1 March he wrote: 'I said to him, "Why don't you give him [Mergenthaler] a contract for 200 machines." He said you wanted too much money. I told him you had told me you would agree to take an order for 200 at $900. He inquired, "Without space bands and matrices." I said, "Of course." He said, "That would make it $975." I said, "What of it. We are pressed for machines." . . . I think you will get the order.'

On 5 March when Clephane cornered Dodge about the contract, he said that Mills was opposed to paying more than $900. [It was reported that the company sold the machine for $3,000 at that time.] Clephane promised to 'go for Mills' about a week later.

Dodge was rattled when Rogers filed an infringement suit against the use of the space band in a federal circuit court but Judge Acheson denied it at the preliminary hearing in November 1893. He acknowledged that the Patent

Office had favoured Schuckers over Mergenthaler but stressed that a judicial determination on the space band infringement issue would have to consider the 'state of the art' at the time the mechanism was invented. As a further ground against issuing even a preliminary injunction Acheson wrote that such a ban against the Mergenthaler company 'would cause serious injury' in view of its established business. This denial of the injunction would cause 'no irreparable damage' to Rogers because he was not making a machine using the Schuckers patent. On 3 December 1894 Mergenthaler wrote in his diary: 'Received news by Mr Dodge of favourable decision in the Rogers Appeal case.'

Rogers may have sensed that it would be futile to pursue the injunction for infringement any further in a federal court. In mid-1894, having failed to make a general case against the space band, the Rogers people started to use the same tactics on Linotype users as Linotype had used on Typograph users in 1890. They issued injunctions against two Detroit newspapers for using the space band on their Linotypes. This action created uncertainty among users and prospective users of the Linotype. According to Dodge the Mergenthaler Linotype Company bought out the Typograph Company in mid-1895 for $416,000 to avoid the disruption and cost of having to make major modifications to the justification mechanism on some 2,500 installed machines.

Of course Mergenthaler was outraged at this settlement, particularly as he considered that his step justifier had solved the justification problem. His anger showed that he did not understand the business decisions behind the transaction. The sale included the Rogers patents, plant, and machinery and thereby eliminated a source of competition. Finally, in a remarkable example of poacher turned gamekeeper, Rogers was appointed consulting engineer to the Mergenthaler Linotype Company and for nearly forty years served it so effectively that some commentators did not realise that he had originally been a competitor. Mergenthaler was always scornful of Rogers and it is not known whether they ever met. However Rogers, who knew Mergenthaler's son Herman, was furious when the *Brooklyn Eagle* printed an article in the Sunday supplement of 5 September 1926 with the heading 'Rogers, Father of the Linotype.' He had them print a retraction next day and insert disclaimers in the trade press. In his apology to Herman he wrote: 'I have an illustrated lecture on the Linotype which I have delivered more than one hundred times all over the United States, and in every case I show your father's picture and give him the honor due, and in every article which I have written in reference to the history of the Linotype I have done the same ... I am writing this letter because you are the only living representative of your father whom it is my pleasure to know.'

## 2  *A short digression about justification*

More than 60 years after the Schuckers application was given precedence over Mergenthaler's it became clear that both devices infringed a patent granted to Merrit Gally. An article, on page 15 of the August-September, 1953 issue of the *L&M News*, cites claim 5 of a patent granted to Gally on 16 July 1872: 'the method ... for increasing the spaces in the line – ie by the simultaneous movement of the compound spaces – to insure perfect and artistic justification.' In 1874, D. B. Ray had also patented a system of double- wedge justification. The existence of these two patents implied that the Schuckers invention had been fully anticipated and that both his and Merganthaler's applications for a patent should have been refused. Both the previous patents having expired by 1892, it was ironic that Mergenthaler paid Gally for the right to use the single wedge spacer on the second band machine, which was not exploited, and then got involved in legal wrangling over an invention that was (albeit unknowingly) freely available!

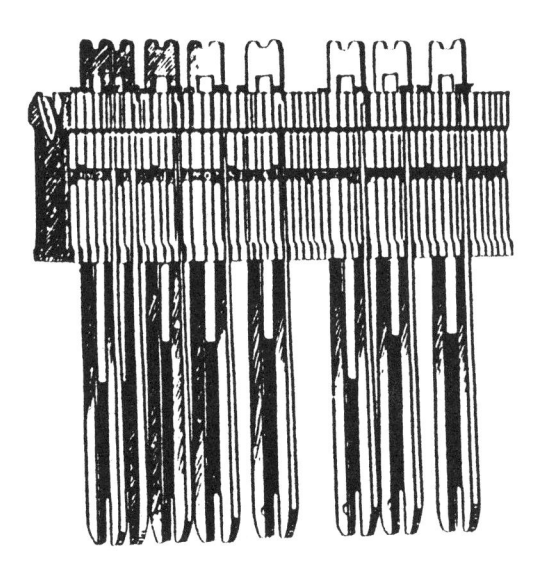

Figure 17   Line of matrices with double wedge space bands
*Source: Iles, Leading American Inventors, p 396*

The Linotype depended on space bands, like those shown in figure 17, to provide variable width spacing between words to spread the line to its required width. The double wedges were designed so that the surfaces in contact with the matrices were always vertical. When the line of matrices and spaces was locked up for casting there were no gaps through which molten metal could escape. The advantage of this system was that the bands widened gradually and continuously as the long wedge was raised.

Ignorant of the fact that Schuckers should have been refused a patent Mergenthaler designed the justification mechanism based on single piece space bands, shown in figure 18. This mechanism went through at least three development phases. Mergenthaler made notes of two versions in his diary with the text and sketches shown in figures 19 and 20. His note (with corrected spelling) of 21 January 1893 read: 'Conceived the idea of justifying the lines by step movement so arranged that while the line is passing over the intermediate delivery channel the space bands are each advanced one eighth of an inch more than its predecessor. After going down into the vice a lever is to move them further until the line is closed out after which another bar comes up and brings the spaces that have entered the line high enough to close the mould.' For the

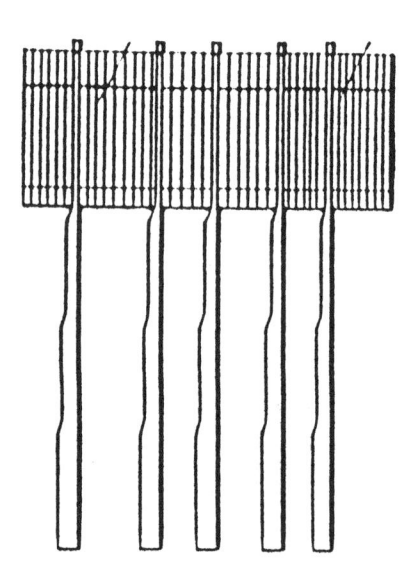

Figure 18   Mergenthaler's graduated wedge justifier
*Source: Iles, Leading American Inventors, p 414*

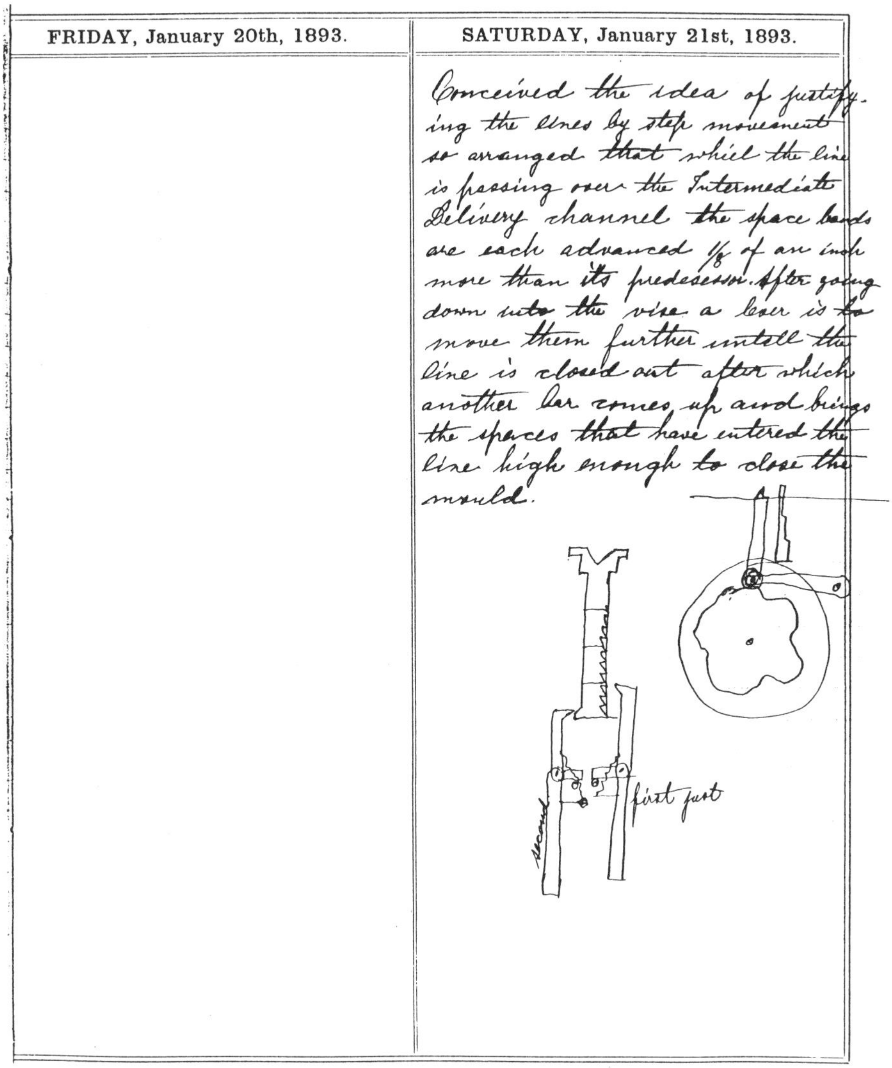

| FRIDAY, January 20th, 1893. | SATURDAY, January 21st, 1893. |
|---|---|
|  | Conceived the idea of justifying the lines by step movement so arranged that while the line is passing over the Intermediate Delivery channel the space bands are each advanced 1/8 of an inch more than its predecessor. After going down into the vise a bar is to move them further untill the line is closed out after which another bar comes up and brings the spaces that have entered the line high enough to close the mould. |

Figure 19   Copy of Mergenthaler's diary for 21 January 1893
*Source: Mergenthaler family papers*

WEDNESDAY, October 18th, 1893.

[Sketch with labels: side elevator, spring, rack, pawl, top view, space bar]

THURSDAY, October 19th, 1893.

Invented the way of inserting or rather stepping the justifying spaces without the use of the cams and gears now used for that purpose on Ref. justification No. 1. It consists essentially of making the heads of the spaces to be elevated by traveling up an inclined surface while in the act of transfere from the setting to the casting position. This incline is supported by a rack of the same endless as it used to raise the spaces in elevator and a pawl, which is detented by the passing space bands allows the ~~rack to~~ inclined surface to drop one tooth as often as a space passes it, thus dividing the spaces between the different teeth of rack in Elevator. See sketch. The inclined surface to be returned to original position again by the "delivery" or some other motion on the machine.

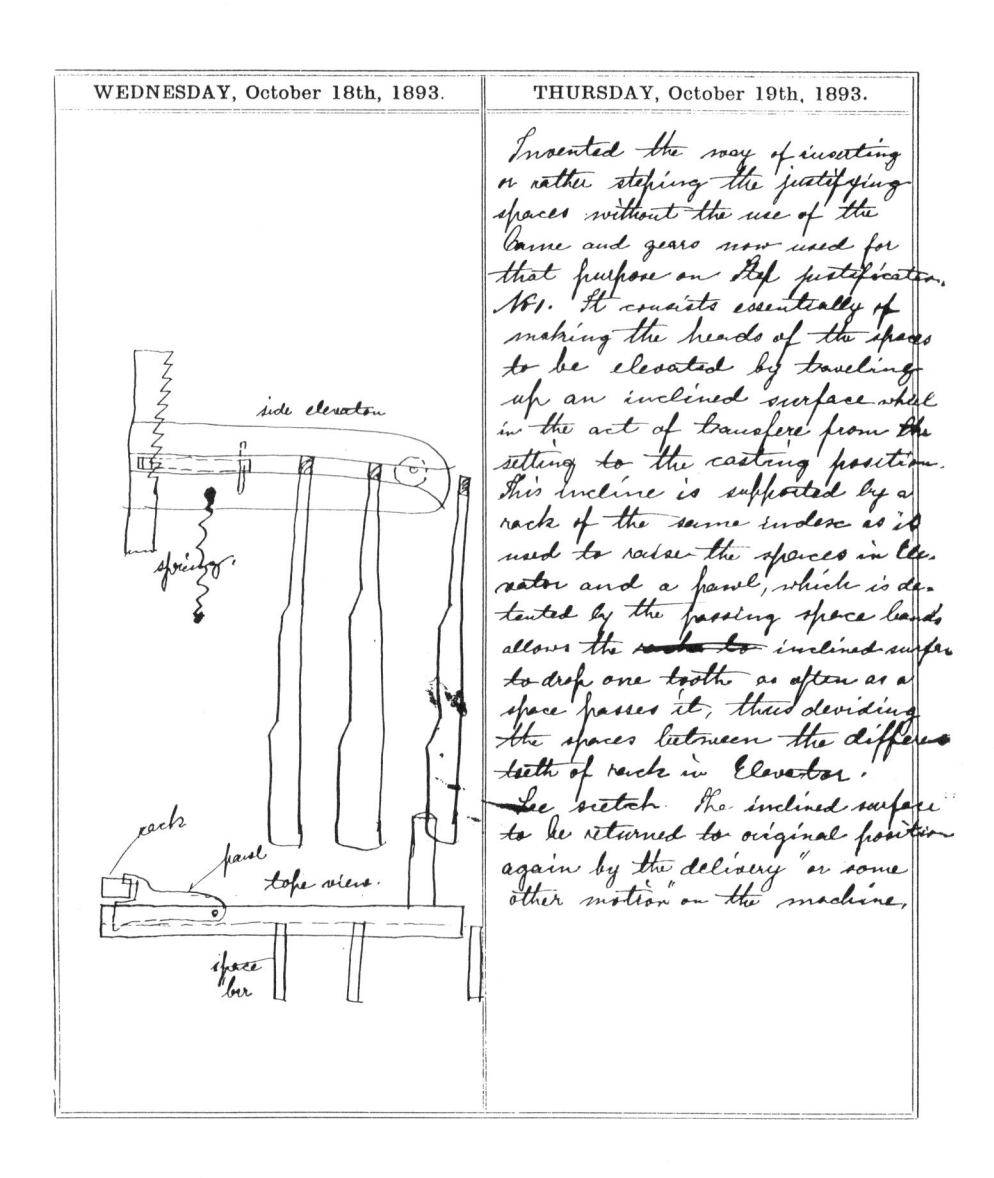

Figure 20  Copy of Mergenthaler's diary for 18 October 1893
*Source: Mergenthaler family papers*

upgrade on 19 October 1893 he wrote: 'Invented the way of inserting or rather stepping the justifying spaces without the use of the cam and gears now used for that purpose on step justification No 1. It consists essentially of making the heads of the spaces to be elevated by travelling up an inclined surface while in the act of transfer from the setting to the casting position. This incline is supported by a rack of the same index as is used to raise the spaces in elevator and a pawl, which is detented by the passing space bands allows the inclined surface to drop one tooth as often as a space passes it, thus dividing the spaces between the different teeth of rack in elevator. See sketch. The inclined surface to be returned to original position again by the delivery "or some other motion" on the machine.' This basic form of discrete spacing was an operable device, but not as satisfactory as the double wedge. It had two major drawbacks: first, it was more complicated than the double wedge mechanism, the single wedges were first driven and then pulled upward by an auxiliary lifting movement operating on their ears; second, the matrix walls would be worn by forcing the stepped bands upwards to spread the line. This mechanism was dropped when the company acquired the Schuckers patent. Dodge's mechanism does not seem to have been successful.

## 3  The Monoline

The Monoline, shown in figure 21, was a single operator linecasting machine with automatic justification and distribution invented by Wilbur Scudder. It was an obvious descendent of Mergenthaler's second band machine but Scudder made it more flexible and compact by replacing the single long bands with eight small bands, each carrying characters of about the same width. The modifications and new features were novel, but not sufficiently novel to challenge the Linotype patents. The Monoline was no match for the Linotype in the field of newspaper production; it was slower and less adaptable, but the company had to protect itself against the potential threat to its sales to the producers of local weekly papers, magazines and books, and to jobbing printers.

It is all but impossible to give a coherent account of the development of the Monoline because odd facts from a variety of sources suggest that Scudder, who was sponsored by Hine among others, was developing this competitive product in company time with company resources.

Dodge must have been aware of the position when he took over from Hine, yet he allowed Scudder to remain as Superintendent of the Brooklyn works until he resigned of his own accord in June 1892. Other directors must have known about developments of which Mergenthaler was ignorant. He had recruited Scudder in Baltimore and always treated him with respect, in contrast to his remarks about Rogers and the Typograph. In the biography he said: 'It is not our intention to discredit Mr Scudder in any way form or manner, but only to point out the fact that no ordinary man is ever able to render efficient services to anybody with his monetary interests opposed to the interests of his employer, and that in allowing this man to continue in charge of the company's manufacturing establishment Mr Dodge engaged in an experiment with all the chances of success against the company.'

In 1891, Mergenthaler had invented a machine similar to the Monoline, but was not allowed to develop it, presumably under Hine's orders. When he applied for a patent the Patent Office declared an interference between Scudder's application and his. On 26 January 1894, Clephane wrote to Mergenthaler, in

Figure 21   The Monoline
*Source: Thompson, History of Composing Machines, p 111*

confidence, that in his preliminary statement Scudder had sworn that he had conceived and communicated his invention of the matrix containing multiple characters of same thickness, to different parties in May 1889; while Mergenthaler claimed to have done so in October 1890. Clephane hoped that when Mergenthaler testified he would not be as modest as he generally was, and would meet the other fellows in the same way as they met him. He hoped that Mergenthaler would prepare himself to squelch them and asked: 'Can't you go further back than Oct 1890?'; but apparently the inventor was too honest to perjure himself.

Two days later Clephane wrote that Dodge had told him nothing about the Monoline case and that all his information came from Rogers so he asked Mergenthaler to treat it as confidential. [This was Rob Rogers, not the inventor of the Typograph; presumably he was a company attorney.]

Mergenthaler wrote in his diary on 2 February 1894 at the end of the examination of Scudder that Dodge had said: 'The only thing against us is that they made a machine while we made none . . . the practice of the Patent Office was to favour the party which developed its inventions.' In the biography Mergenthaler commented: 'A stronger impeachment of his course in this case could not be drawn up by his most severe critics as is implied in the above admission. Here we have the President and General Manager with almost unlimited power and means sitting calmly by and seeing intending competitors develop a machine which he knew from the very day he assumed his duties as manager was supposed to be really the invention of Mr Mergenthaler, and thus enabling the Monoline people to get a standing in the Patent Office which they never would have had if he had given authority to develop Mr Mergenthaler's machine.'

Mergenthaler reported in the biography that the Monoline had made considerable progress; several machines had been built and one had been shown at the Chicago Exposition. This was the outstanding event of 1893 largely devoted to mechanical inventions including printing machinery. Mergenthaler noted in his diary on 20 June that he left for Chicago with his wife Emma, his sister-in-law Anna and his half-brother Fritz. It is not known whether he saw the Monoline; he made no comment to show that he had actually examined the machine.

Opinions about the Monoline varied from enthusiasm to indifference. For a while it looked as though the machine would be a great success. On 10 October 1893 Clephane wrote that he had just received a Monoline circular from the

[Chicago] fair grounds. The machine certainly looked simple and was offered for sale at $1,000 or $250 a year rental. Two weeks later he wrote that it was rumoured in New York that it had been awarded first prize at the Chicago exposition, which Clephane thought suspicious.

Dodge went to see the improved Monoline in Canada that September and told Mills that he did not think much of it and that if they owned it they would not want to manufacture it. Fowler, the inventor of a new press, reported that slugs produced by the Monoline in Chicago had very imperfect alignment; he claimed that even if the Monoline showed good alignment at first it could not last. George L. Bradley (one of the directors) said that although he went two or three times he failed to see the Monoline working at Chicago.

Dodge called on Mergenthaler at about 4 pm on 31 October. He had seen the Scudder machine in Chicago and admitted that it was neat but did not attach much importance to it. He would seek an injunction at the first opportunity and was satisfied that the company could stop the Monoline under Linotype patents. He did not mention the two cases which were in interference with Scudder.

A measure of the way in which Scudder had used Mergenthaler's ideas and company resources can be gained from the following two extracts. On 12 December Clephane wrote: 'By-the-way Scudder freely admitted to a party that his casting device was precisely similar to that in the old *Tribune* machines. Is it possible that Dodge has failed to cover that properly with a patent?' and on 5 January 1894 Mergenthaler wrote in his diary: 'Mr Sharf, in charge of the Canadian [Linotype] Co's shop calls and is shown through the factory by me. He admits that the Scudder machine was practically built by the Canadian Co and in their shop.'

Apparently the company tried to buy Monoline out to avoid litigation, but the asking price was so high, presumably because of the award of the prize at Chicago, that they went to court instead. Both the Typograph and the Monoline were ultimately banned in the United States of America and the United Kingdom, but were manufactured and sold, particularly in Canada and Germany where the Linotype patents did not apply.

It is often thought that travel was slow in the late nineteenth century, but Mergenthaler wrote in his diary on 3 November 1893: 'Dodge sends preliminary statement of second interference with Scudder No 16,270 for my signature. His man carrying the papers arrives here 1.45 pm while the papers must be filed by 4 pm at Patent Office in Washington as stated by his messenger... Messenger left me at the corner of St Paul and Fayette street at 2.20 pm.

Train leaves 2.30 arriving at Washington at 3.20.' This was really cutting things fine; could a messenger meet this schedule on public transport in the late twentieth century?

## 4  Confrontation with Dodge

Mergenthaler wrote that personal relations with Dodge when he took over were polite and cordial despite disagreements over payments and policy. This formal civility was observed in entries to Mergenthaler's diary; even when attacking Dodge he usually referred to him as Mr Dodge. The first argument arose shortly after Dodge became general manager and asked Mergenthaler for a statement of all the money owed to him. It came to $1,300 for patent drawings prepared by Mergenthaler and his assistants for the period from April to December 1888 after he had left the company. This bill had been on the books for four years but Dodge declined to pay that rather modest amount on the grounds that there had been a change of clerks; the new clerk had failed to find this old charge, the company had never given an order for the work to be done and 'that it was not a legal charge against the company.' Mergenthaler was particularly annoyed about this debt because he had 'used his time and money in the interest of the company at a time when the latter in a blind frenzy tried to ruin him financially.' There had been a substantial increase in the income of the company as a result of this work and the claim was 'full of merit and equity.' The inventor continued to present the bill once or twice a year until it was finally paid in full about four years later.

When Scudder resigned Dodge asked Mergenthaler to advise him on the appointment of a new superintendent. The inventor offered the services of his own superintendent since the Baltimore works could make machines profitably for $1,050 whereas Brooklyn machines cost nearly twice as much. Dodge decided to appoint men with no previous experience of Linotypes, although they became effective later. Of course it is possible that Dodge, being a devious character, suspected that Mergenthaler had an ulterior motive in offering to second his own man to Brooklyn.

Further contention arose because of the success of the product. There were insufficient skilled operators and maintenance mechanics to handle the installation of 200 new machines in six months. There were numerous breakages and heavy repair bills, largely due to inexperience. Dodge was concerned that the cost of repairs could harm the business. Mergenthaler proposed modifications to reduce the chance of accidental damage and pointed out that with experience the number of breakages would drop. The company bore responsibility for high repair costs because some spare parts were sold at 400% profit. In the biography he claimed: 'With characteristic indecision, the new president neither adopts the improvements suggested by the inventor nor does he reduce the cost of repair parts, but calmly continues to bombard the inventor with letters of complaint.'

On 26 January 1893 Mergenthaler recorded in his diary; 'Left for Chattanooga to investigate the charges of bad workmanship brought by Knight, the machinist in charge.' The next day he wrote: 'Arrived at Chattanooga and find the machinist incompetent. A few hours work set the machines all right.' Three days later Knight volunteered the information that Randall (General Superintendent of the Brooklyn factory, see figure 22) had asked him to make to make a full report of what he thought about 'those Balto machines' and that he had been told to report them as unfit for work; just one example of the methods used by head office to discredit Mergenthaler's products.

Clephane warned the inventor that Dodge was always trying to belittle the products from Baltimore. On 12 December 1893 when a metal pot developed a leak that could not be fixed he asked Mergenthaler to send a replacement and commented: 'I might get one at the factory, but I would prefer not saying anything to Dodge about it for you know how ready he is to find fault with Balto.' On 17 April 1894 Clephane wrote: 'Mills stated that you had stated that you could not make the parts interchangeable with Brooklyn machines. B [Bradley or Bryan] & I denied this, and this proviso was inserted. D[odge] undertook to disparage your justifier. I upheld it. He pitched into your machines, but I floored him on that. I asked him to point to any office where your machines were not giving satisfaction. He referred to the Chattanooga office, I believe. I told him I could not say as to that, but that I knew *all* the Phila parties & the Balto parties, and all other parties I knew of insisted on having *your* machines. He then said they were not *durable*. I said that was a matter that I thought nobody was competent to speak of at the present time. I floored him, however, in this and several other matters. I suppose he will seek to have me off the Board next time, but I was determined to fight for you.'

A drawback of Linotype setting was that matrices wore out, leading to the formation of hairlines [burrs] which made it unsuitable for high quality book production. Mergenthaler tried to overcome this by making matrices of hardened steel. (See chapter 8.) He demonstrated his experimental steel

# DODGE'S RULE AND CONTENTION FROM OTHER LINECASTERS 111

Figure 22  Top: Letterhead showing Simplex Linotype and factory
Bottom: Picture of Brooklyn factory from back of Dodge's 1893 report

matrices to Dodge in December 1893 and was very disappointed when the manager showed no enthusiasm and said as usual: 'Yes, if it can be done it ought to be done.' However, Dodge vetoed production of steel matrices a few days later, because of its impact on the company's business with brass matrices which were protected by patents. Dodge may also have budgeted for an income generated by the need to replace brass matrices regularly.

Dodge was named as attorney on all but four of the inventor's patents. By the beginning of 1893 it was evident that Mergenthaler was becoming disillusioned with Dodge's management style and his handling of patent applications. Although he was the submitting attorney, Dodge must have delegated much of the clerical work after he became President and General Manager. As an active inventor who made several major enhancements to the Linotype he had to keep abreast of Mergenthaler's work.

The inventor went to New York at Dodge's request on 7 February 1893. The next day (the day his father died) they discussed inventions that had been submitted to the manager to apply for patent before Mergenthaler had left for Europe the previous summer. The inventor wrote in his diary: 'He had some of the patent drawings there but evidently had never looked at them before nor at the case as a whole. He appeared to be absolutely ignorant of it.' Mergenthaler was infuriated by this incompetence. The next day he complained about Dodge's management to directors Darius Ogden Mills and his son Ogden Mills. He had to make his complaint but it was unlikely to succeed because Dodge was Mills's nominee.

On 13 February Mergenthaler told Dodge that documentation of the new versions of the justifier was ready to be submitted for patent, but having heard nothing for a month, the inventor wrote about the delay on 15 March. The applications arrived for signature on 30 March but were full of errors and had to be returned for revision. Mergenthaler wrote in his diary: 'The misstatements contained therein were so glaring and the cases contradictory in themselves that it is hard to see how such a thing could happen unless it was done on purpose.' He sent Dodge documents for the improved justification with a full set of drawings on 5 April. Dodge came to discuss patent matters with Mergenthaler on 11 April. He had not filed the applications for any of the justification mechanisms and asked for two more diagrams to go with the documentation sent on 5 April. He claimed never to have seen or known of the description supplied with the first justification case dated 23 July 1892 and asked for a copy. Mergenthaler underlined this entry in his diary: 'That same copy was furnished to him a second time March 30 with the returned cases.' Dodge had not read papers submitted on 5 April and claimed that he had submitted other cases at least a year and a half before. Mergenthaler showed him the new justification working well but Dodge did not 'utter a word as to the possible effect on the Schuckers case.' The inventor wrote: 'My impression is that he would rather see the same a failure than a success.' Mergenthaler showed Dodge a third form of justification and other devices on 29 November 1893.

On 10 January 1894, the inventor met Bryan, one of the directors, at the Telephone Exchange to discuss company affairs, the failure to pay his bills and his suspicions about Dodge's handling of the company's patent business. He wrote in his diary: 'Received copies of patents appearing on "Patent plate" on the machine and find that Dodge had applied for a patent on Nov 11, 1889, the same day on which my own application for substantially the same thing went into the office. Claims in his patent are very broad and in mine correspondingly limited. There is another patent out by Dodge filed the same day on a compressible step space band which was issued Jan 6, 91, Number 444,337. At least two of my applications covering the same general object are still unacted upon although filed earlier.' A week later he submitted an outline report to Bryan followed by a full statement on 14 February.

On Tuesday 19 June 1894 Mergenthaler wrote in his diary that he had completed successful experiments with a two-line [drop] letter used for setting small advertisements in newspapers. Although Matthew Whittaker had invented a similar device in England in 1893 there was no contention over these inventions because the British and American companies had agreed not to compete against each other.

Mergenthaler had become so seriously ill by mid-1894 that he gave up active management of his works to consult specialists in search of a cure. He tried to keep abreast of the business but was mistaken over details. He thought that the English interests had been sold in 1894; in fact this was when Hutchins was finally paid $280,000 commission, five years after he had negotiated the sale. In the biography, Mergenthaler accused Dodge of trying to award himself an enormous cut from royalties payable on the sale of manufacturing rights to a German company, but the Board struck out the clause dealing with that payment.

Mergenthaler finally broke with Dodge over a letter dated 19 October 1895. The manager claimed that the company name was too long and that it was often misspelt and suggested shortening it to the Linotype Company and implied that he was just asking the inventor to agree. Mergenthaler was deeply hurt by the suggestion and noted in his reply of 23 October: 'The company has borne my name during periods when it bought me more ridicule than honour and more aspersions than credit . . . the machines are referred to quite as frequently as "Mergenthalers" than as Linotypes.' [This persisted after the inventor's death; in 1903 the technical writer John Thompson referred to the Linotype as the Mergenthaler in the Convention Trade Book of the ITU.] Nevertheless, Dodge persisted with his plan to change the name of the company, so the inventor appealed to D. O. Mills, who, for the first time, decided against the manager.

Mergenthaler's bitterness towards the company and its management may well have been aggravated by his illness. On 21 April 1896 he wrote in his diary: 'Was in New York at the meeting of the Board of the M L Co and presented charges of misconduct in the management of my patent cases against Mr Dodge. Also made various other charges of misconduct against him. No action by the Board.' This was to be expected; the Board was not likely to incriminate itself by upholding his complaint.

On 8 May 1896 he wrote: 'Was in Washington attending a meeting of the Board of Nat T Co. Present; Dodge, Mills, Murphy, Devine, Clephane & Stilson & Lee Hutchins. Gave notice that I demand 10% of the cost of making all machines in foreign countries outside of England. Ogden Mills passed a resolution for the appointment of a committee to find out whether my claim had any foundation under the existing agreements. Other business, German sale and Patent matters.' [The National Typographic company had remained a separate entity even though Hine had suggested that it would become part of the Mergenthaler Linotype Company.]

Dodge envied Mergenthaler his creativity and his royalty. The manager was paid $10,000 a year and obviously hated to think that as production increased Mergenthaler was receiving royalties of $50,000 a year as well as his income from manufacturing Linotypes.

## 5   *Mergenthaler and Clephane*

Many of Clephane's letters to Mergenthaler written between 13 September 1893 and 29 June 1894 have survived. They show how close the two men became. On 1 January 1894, Clephane dropped formality by writing 'Dear Mr Merg' and later went so far as to address the inventor as Mergenthaler or just Merg. These letters indicate that Clephane acted as Mergenthaler's eyes in New York and that he put himself out to represent the inventor's interests before Dodge and the Board.

## DODGE'S RULE AND CONTENTION FROM OTHER LINECASTERS

By 1893 Clephane had been pouring money into the development of mechanised typesetting for about 27 years and had no further resources. An associate had let him down badly and he wrote to Mergenthaler about his troubles. Unfortunately there is no record of the inventor's reply but Clephane was deeply touched; on 11 December 1893 he wrote:

> Your letter affected me to tears. I have been so harshly treated by every body here connected with the Linotype, notwithstanding the time and money I have spent on the enterprise, that your letter was like the weary and thirsty traveller coming across an oasis in the desert. It is the first kind and appreciative word that I have had. Dodge is so jealous of every one who had anything to do with the enterprise, that I get no sympathy or encouragement from him. When I told him of my misfortune his only reply was a smile, and you know that 'smile'. He is the most heartless, selfish and cold blooded fellow I know of. He is now thoroughly jealous of Betts [the lawyer who represented the company in their action against the Typograph] and is delaying the paying of his bill. He claims that he did nine-tenths of the work and ought to get the credit for it. What do you think of this. I hope Betts may hear of it and report to Whitney [a director on the executive committee of the company]. I trust you, Greenleaf and I may get even with him some day.
>
> You are very kind to offer to aid me financially, and I assure you I fully appreciate it, but I hope this may not be necessary. My disappointments and losses have been great, but I will endeavour to bear with them, if I can. I hope you will be patient with me, however, about the payment of the bill. You know I will arrange as soon as I get on my feet, which I hope will be sometime.
>
> I hope you see a way to charge the *experimenting* expense to the Co. I don't see why either you or I should bear that. Can't you make it in some general way, so that D[odge] will not connect me with the charge for you know his feeling for G[reenleaf] self and you. I will insist that it be paid, if you include it in your general bill, even if it should be necessary to go to the Board, which I am sure will not be necessary. For the specific article I, of course, am willing to pay, but the Co ought to bear any ween over the regular commercial price. Don't you think so? . . . Your letter has given me fresh courage.

Obviously there was no love lost between Clephane and Dodge and Clephane paid out the manager in his own way. In his annual report dated 4 October 1893 Dodge noted that the company had paid no dividends but implied that the financial position was so good that they should be able to start paying dividends the following year. Both Dodge and Mergenthaler wanted to buy shares in the company and Clephane made sure that the inventor was offered any that became available. Some stockholders were not prepared to sell their holdings to Dodge.

On 1 March 1894 Clephane informed Mergenthaler that he had seen Dodge that morning. He was in the best of humour and said that he had taken out 75 patents, and some of them valuable, all of which he proposed to turn over to the Co without reward. [This dig at Mergenthaler was fatuous because all his past and future patents had been assigned to the National Typographic Company under his 1884 contract.] Clephane said that was good. Dodge wished he had more time to give to watching the state of the art. He expected the enterprise to be valuable and would like to get stock at $50 but had not been able to secure any yet.

On 27 March Clephane asked about Mergenthaler's new justifier. He was anxious about it, as the machines at the *Herald* and the *World* were beginning to burr, and he hoped that the new justification would make hairlines almost impossible. [This was not the solution, see chapter 8.] On 2 April he advised Mergenthaler to get his own attorney to look after his patent applications, and trust to the company reimbursing him afterwards. Clephane suggested that the inventor should patent a number of devices that he had made for him and continued: 'Our friend [Dodge] is too filled up with jealousy to do anybody justice.' Clephane also reported that the *News* office was very pleased with an automatic feeder for bars into the metal pot got up by a man named Grigg, but that Dodge took out a patent for identically the same thing in his own name. The next day Clephane referred Mergenthaler to Capt W. W. Wood, of Washington, as the most reliable and experienced patent attorney of which he knew.

On 14 June he informed the inventor that the Linotype was fast losing ground to typesetting machinery in the book field because of hairlines and a lack of italics and proper fonts. A leading typefounder, who admitted that the density of the steel would give much better results, thought that the Linotype was at its limit and that users had been talking to him about going back to movable type on account of lack of cleanliness. He said that Dodge admitted to him that the Linotype could never hope to do real good work and that it had reached its limit. Clephane continued: 'I showed the typefounder some of our book work, and Greenleaf's, but he would not believe that was not done from selected bars etc. Now, I hope in order that we may shut one line out that you will press forward with your steel matrices. Such, (there being no burring) will enable us to capture the book field just as soon as our friend comes to his senses.'

In June 1894 the company declared a dividend of one and a half per cent to be paid in August. Clephane wrote that Dodge was apparently considerably disgruntled that he had not been able to get more stock. He informed Mergenthaler that Dodge was looking forward to the new justifier with great interest and felt that if it worked all right, then it would be a comparatively easy matter to deal with the Schuckers people.

Mergenthaler seems to have become very despondent about this time, probably due to increasing ill health, and Clephane tried to jolt him into a more optimistic frame of mind. On 23 June he wrote:

> Now, the idea of a young man like yourself talking about being too old for 'keeping pace'. Why, you are just in your prime, and with money now coming in regularly, you are just in a position to tell everybody else in our line to take a back seat and stay there. Edison accomplished nothing until he was long after your age. No, it is on you that we will have to rely for squelching others by improvements and I know you can do it. Shove forward your steel matrices and then the typefounders, with their typesetting machines will have to take a back seat. You must now keep the laurels you have won.

He continued in the same vein on 4 July:

> I am surprised that you should feel discouraged about improvements, but *cannot afford to be*. Reputation, is a thing, though sought after by many is attained by only a few. By your transcendent abilities you have won it; now you must not by reason of jealousy of two or three, temporarily in authority, permit a lack of effort to in any way tarnish, or diminish it. I have in mind myself two or three schemes for bringing this about. I feel when you have gotten your steel matrices, and your new justification, and thus made burrs impossible, you will have reached the acme, but still there are others actively in the field, and they must not be allowed to step in ahead of the Lino.

He noted that some New York papers were considering using the Empire typesetter to run alongside the Mergs [Linotypes] and criticised Dodge's defeatism with the comment: 'Our friend is not equal to the occasion, and we outsiders will have to do all we can to make up for his short comings.'

Finally, Mergenthaler showed his regard for Clephane by sending him a valuable present. Clephane was the source of both letters so resolving the dates is left as an exercise for the reader. The first letter from Mergenthaler to Jas O. Clephane, Esq, dated 18 July 1894, from Baltimore, Md, read:

Dear Sir, It is now nearly 18 years since I had the good fortune to make your acquaintance. I say 'good fortune,' because I realise that it was you and your untiring devotion which started the invention of the Linotype, and kept it alive during the many years of labour and adverse conditions until at last we succeeded in bringing it before the public as a commercial success. During all those many years I found you an unfaltering and devoted friend; one who never left me, in appreciation of which I have sent you today a slight token of friendship and admiration. I hope it suits your taste.
With my best wishes for your future happiness and success, I remain,
Yours very sincerely,   Ott. Mergenthaler

Clephane's reply was dated 17 July 1894, from Englewood Cliffs, New Jersey:

Dear Mr Mergenthaler,
On arriving at my Englewood home I found awaiting me a most beautiful gift from you, and at the same time your very complimentary letter explaining the occasion of its sending. While the present itself is of a character to call for the most profound thanks from the recipient, and one to be most highly appreciated for its great intrinsic value, yet the very pleasant words contained in your letter, I assure you, are received with even greater satisfaction and pleasure.

It is so seldom in this world one finds a recognition of favour, small, or great, that it is like a lovely oasis in the desert of life, to meet with such tangible proof of appreciation of what one has at least tried to do, although his efforts have been humble, and, too often ineffective.

I shall prize these volumes above all other gifts, and, preserved in a permanent form, shall go with them the kind impressions indited by you.

And not satisfied with this act of kindness, you have been disposed to add another, the cancellation of my account. This certainly puts me under great obligation, and I trust I shall soon have it in my power to reciprocate in some substantial way, these gracious acts on your part.

As before, I shall ever be on the alert to give information – for this is all I can do – which may have the effect of making the Linotype, your offspring, a thing to be unsurpassed and which will, for all time to come, be a perpetual monument to your genius, your industry, and your sound judgement. Please accept my most sincere thanks, and believe me ever,
Your faithful friend,    Jas O. Clephane.

# ❦ SIX ❦

# *Mergenthaler*

## 1  *The domestic picture*

Mergenthaler was no absent-minded professor type in an ivory tower. He was a family man who loved his children, although he worked such long hours that there must have been times when he hardly saw them. When he had spare time he enjoyed singing with his friends at the Liederkrantz, of which he later became president, and liked to drink, smoke cigars and joke with his fellows. His daughter, Pauline, who was only five years old when her father died, remembered riding down a block or two in a horse-drawn carriage with him in the mornings as he drove off to work, but she never explained how she got back up the hill by herself at that early age. She described her father as being extremely dedicated to his work, serious, heroic, noble. Her brother Herman, who was some seven years her senior, remembered him a little differently – dedicated to his work, yes, but with a sense of humour, a man who enjoyed a drink with his friends. His wife, Emma, would share family jokes with him. When her sister complained that her life lacked excitement Ottmar wrote to her proposing a clandestine rendezvous – this letter bore a facetious comment from Emma.

In his youth Mergenthaler had been very fit but, within ten years of meeting Clephane, he had ruined his health by overwork and dedication to his invention. Yet, apart from the near fatal bout of pleurisy in 1888 after his resignation from the company, there was no mention of his health in the biography until 1894 when the symptoms had become so critical that he gave up active management of his factory. However, others had been concerned about the inventor's health both before and after the attack of pleurisy. On 22 June 1886, shortly before the first machine was installed at the *New York Tribune*, Smith informed Reid that Mergenthaler was so ill with lung trouble that he might have a complete break down. Later, on 24 November 1890, Hine wrote: 'I regret that you are not in the best of health. It will not pay to use that up and worry will do it faster than work.' The inventor probably believed in his own immortality and persisted where a less dedicated man would have withdrawn from the project.

There had been five children in the family. Four boys: Fritz Lillian, born in 1883; Julius Ottmar, 1884; Eugene George, 1885; Herman Charles, 1887; and Pauline Rosalie in 1894. Apart from Julius, who died in 1888 during one of the most traumatic periods of Mergenthaler's life, they all survived their father and received the higher education that he lacked. Unfortunately, the line did not endure.

Fritz, who attended Cornell University and became a mechanical engineer in Baltimore, married Doris Feldner in 1909. The next year he, his wife, her parents and their chauffeur were killed instantly when their car was struck by an express train in New Jersey at 5.47 pm on 9 August 1910. The car disintegrated and, apart from Mrs Mergenthaler, its passengers were so mutilated that they could only be identified by their clothing. Next day the *Evening Sun* corrected the account given in that morning's issue of the *Baltimore Sun* to report that Fritz was driving the car at a leisurely pace but it was very noisy. May McNeil, a 20 year-old girl, heard the whistle of the approaching train and tried to warn the driver but he did not notice and the car passengers were looking the other way. To save a couple of miles they had taken a short cut which took them over the unprotected double-track crossing, obscured by tall corn. Mrs Feldner had had a premonition and had tried to dissuade her husband from going, but they 'jollied' her out of it even though she was still anxious when they set out. The morning report included several paragraphs about Ottmar Mergenthaler and the Linotype under the heading 'His famous Father'.

Eugene attended Johns Hopkins and Cornell Universities, became an electrical engineer and ran his own company in Baltimore. He did not marry and died of influenza during the 1919 epidemic.

Herman died in 1972 at the age of 85. His son George Ottmar, born 1920, was killed in action in the Battle of the Ardennes in 1944.

Pauline who married Rody Perkins in 1917 and had a daughter Nancy in 1918, resumed her maiden name after her divorce, and died in 1986. Nancy never married.

The Mergenthaler family was a close knit group and, apart from his singing, the inventor preferred to stay at home. It was said that the family took their vacations together and when he went to Europe in the summer of 1892 he took the family with him. This visit to Germany was a purely domestic trip, he was not invited by any vested interests, and it turned out to be his last chance to see his father who died on 8 February 1893. Therefore he did achieve his main objective – to make his father proud of him. It was a happy family reunion. Ottmar loved German cooking and his sister Caroline plied him with such large portions of his favourite dish, pork with sauerkraut, that Emma told him to curb his appetite or he would make himself ill. Caroline only commented that one is a long time dead. He also went to Bietigheim for a couple of days to look up his step-uncle Louis Hahl, his former master, but the old man had died in 1884 and was buried in that village. Forty years later an old inhabitant remembered that Mergenthaler had climbed the tower to see for himself that the clock that he had restored all those years before was still working. It had been moved to Bietigheim when the Ensingen clock tower was demolished.

## 2  *Relations with his men*

Mergenthaler was a very loyal character, it was said that his friends were friends for life; unfortunately the corollary was also true – he had a short fuse and was thin-skinned so he could also be an implacable foe.

Davids reported problems with pro- and anti-Mergenthaler factions when he took over the Camden and Preston Street factories in Baltimore, but much of this could have arisen because members of the Board had interfered with Mergenthaler's management. In general the inventor got on well with his workers, probably because he had served his time on the shop floor and understood their problems. In a short biography, in his book *Leading American Inventors,* Iles wrote that Mergenthaler had the personality which makes an employer beloved by his hands. His men were proud and fond of him. They rendered him ungrudging service; their good will did much to cushion the jolts of experimental work with its inevitable hitches and its constant balking of the best laid plans. A member of his staff, William R. Brack, declared (in politically incorrect language) that Ottmar Mergenthaler was the 'whitest' man one could work for. He was good to his employees, and no matter how humble their station, always had a kind word for them, and a friendly word to say of them. His goodness of heart included dumb animals, horses especially, and he would not permit them to be ill-treated. One evening Brack rode with Mergenthaler in the horsecar. At an unpaved crossing the driver lost his temper and began to whip the horses unmercifully. Mergenthaler sprang to their rescue, and gave the driver such a reprimand as he had never heard before. It had the desired effect, that man never abused his horses again!

Charles R. Wagner, of New York, another machinist, who helped to build the first Linotypes said that there never was an employer better liked than Mergenthaler. When rush orders obliged all hands to work overtime, he would walk through the shop and ask if they had dined. If they said 'No', he would send for dinner from a neighbouring restaurant. When Mr Hine resigned as president of the Company, Mergenthaler gave all hands a capital supper at the shop and made a telling speech. [It was reasonable, but unusual, for an employer at that time to feed his men when they had to work overtime; they were obviously paid by the day, not by the hour.]

Mergenthaler's management style, based on encouragement rather than threats, differed radically from that of Reid and particularly Dodge who favoured the 'treat 'em mean and keep 'em keen' approach. The incentive system that was planned to raise productivity infuriated Reid who thought that the men were overpaid. The inventor considered it outrageous when Davids, with the approval of Reid and the syndicate, used a loophole to cancel contracts 'with the avowed purpose of confiscating the money which honest and hardworking men had earned and which belonged to them as much as his own stipulated salary belonged to him.' Mergenthaler was highly strung and resented unrelenting criticism based on purely financial criteria. He pushed himself harder than his men and earned their respect by example; he could do any job in the shop as well as, or better, than any of them. He expected others to pull their weight and would not accept slackness or indifferent work. He was a fair but blunt employer with a lack of diplomatic skills as shown by his handling of complaints about Berger's work in the matrix shop and Johnson's excessive drinking.

The inventor had only a primary education and no training in geometry but he did not try to hide his ignorance and readily accepted the advice of an apprentice. When he was using dividers to count the number of cogs in a wheel (a technique used by eighteenth century wheelwrights to set the spokes in a wheel) Charles Letsch told him how to calculate the number by using the formula $2\pi r$. Mergenthaler did not resent the intrusion; instead he said, 'Good boy, now find someone to do your job and come and be my assistant.'

However, he would not have been human if he had been without faults. When he blamed the draughtsman for the heavy square base he knew that the man tended to make things too bulky; the inventor was responsible for approving the plans. Due to pressure of work he accepted sub-standard castings from contractors instead of demanding immediate replacements to the agreed specification. Further to this, he had told Reid that his invention of the Linotype was merely a means to an end – his real ambition was to be the head of a large commercial enterprise. Presumably he thought that he would appear weak if he called on the support of others when he had a problem; in fact this attitude demonstrated an aspect of weakness.

Mergenthaler expected generous rewards, largely because he was both inventor and builder, but he did not believe in overpaying his men. When he looked for skilled toolmakers he paid from $21 to $27 for a six-day week. At this point there is a paradox about Mergenthaler's industrial relations. When ill health forced him to give up work in 1894 he asked his half-brother Fritz to manage the Locust Point factory. Fritz, having seen how Reid and Dodge had treated Ottmar, did not want to suffer in the same way and asked for more money to accept the responsibility. He was offered stock in the business but his wife, Rose, said: 'You can't eat stock certificates', and in the words of Fritz's grandson, Ronald Mergenthaler, Fritz effectively told Ottmar: 'To shove it!' Maybe both Ottmar and Fritz inherited their father's stubborn streak. This caused a total breach that was never healed. Emma was unhappy about the rift; she thought that Fritz had not appreciated all that Ottmar had done for him. Mergenthaler did not mention Fritz or his family in either his will or his biography. He appointed Carl Muehleisen as superintendent of his factory at $175 per month, considerably less than the $3,000 per year that he had been paid to run the experimental Camden street shop in 1884.

## 3   Comments about his creativity

Edison was reported to have described the Linotype as the eighth wonder of the World and also as one of the ten great inventions of the nineteenth century. Although these comments were inconsistent, it is evident that the great inventor had a high opinion of Mergenthaler's creative abilities.

Harry G. Leland, who described himself as the first Linotype operator, worked for Mergenthaler at Baltimore. He had problems with distribution on the Blower Linotype and recalled that Dodge told Mergenthaler: 'One of these days you will come to a screw feed for the distributor.' This was implemented on the Square Base, but Mergenthaler did not mention that it might have been based on Dodge's suggestion.

The Australian printing historian, Bullen, in a two-part article in the *Inland Printer*, in 1924, wrote him off as a plodder because of the number of unsuccessful machines that he had produced before he achieved success. This is most unfair; these so-called 'unsuccessful machines' were experimental models that pointed the way to an effective product. Either Bullen did not realise, or just ignored, the fact that the early machines were built under contract, using techniques specified by Mergenthaler's sponsors, and gave him no credit for trying to stop his backers from placing large orders for untried machinery. His progress was very quick when compared with development cycles in the late twentieth century, particularly as he was pioneering new technology. From 1 January 1883, when he started work on the first band machine, he took only a year and a half to change the technology and build the first hot metal demonstration machine. Little more than six months later the second band machine with automatic justification, was shown to the public at the Chamberlain Hotel. Less than a year and a half after that the first Blower Linotype with freely circulating matrices was installed at the *New York Tribune* and, by the end of 1886, despite teething troubles, it had been used both for newspaper setting and for producing the 500-page *Tribune Book of Open-Air Sports*. Later, the inventor was continually harassed to stop making changes and get on with production. Bullen did say that the Blower Linotypes at the *New York Tribune* produced good slugs in 1888 and he did much to popularise the machine outside the USA.

Bullen admitted that Mergenthaler was a conscientious man of fine character, but praised Schuckers, for the double wedge space band, and Benton for the punch-cutting machine, more than the inventor for building a practicable composing machine. He ignored the facts that Mergenthaler was refused a patent on the double wedge space band because Schuckers's application had been submitted 49 days earlier (which does not mean that he had copied Schuckers's invention) and that he stopped work on a punch cutter because Dodge had acquired the Benton and Waldo engraver for the company. On the other hand, Letsch, reminiscing about Mergenthaler in 1936 claimed that he was very quick and astute, definitely not a plodder. Letsch spoke from first hand experience of working with the inventor for years while Bullen, who had seen the Blower Linotypes at the *Tribune* in 1888, probably never met him.

In 1954, on the centenary of his birth, Mergenthaler's son Herman wrote:

> Quite apart from Linotype labours, his mind was always concerned with a
> practical application of effort when confronted with problems related to time,

cost and labour saving. I recall a grotesque horse-drawn, earth-moving apparatus. It was not unlike the present-day motorised units; but it served his temporary needs to put through a road at Locust Point, Baltimore, with a minimum of time and cost. For efficiency and speed he built for his own shop use a multiple-speed sliding gear transmission for lathes and milling machines. This mechanism was an extreme innovation, since standard machine tools had the conventional step-belt drives. When this new machine was subsequently offered for general sale, possible purchasers were skeptical and few were sold. Unfortunately for him, this and other of my father's devices were years ahead of general acceptance in his lifetime.

Mergenthaler has been described as 'an inventor through and through.' Although primarily occupied with inventing and improving the Linotype he did not limit himself to it. He also invented a Pneumatic Clock jointly with August Hahl (patented on 17 March 1885 with P. T. Dodge as patent attorney); a threshing machine; and an improved basket-making machine. In April 1894, despite failing health he experimented with a perfecting press (one that prints simultaneously on both sides of the paper); the following month he was working on a machine to plough and grade soil. Barely a year before he died he was preparing the specification of his final composing machine. On 25 October 1898 he wrote in his diary that he had finished the sketches and written the description of a machine with 256 characters, 64 keys and 4 channels for casting single words (logotypes). Like the Linotype, this machine would compose lines of matrices but it would not cast solid lines. Instead it would cast single words with a space attached to justify the line. Mergenthaler claimed that it would give better slugs and make it easier to correct errors (but did not seem to realise that the output would have to be handled like movable type). The detailed description and sketches of the machine covered twenty sheets of closely written letter head paper, many written on both sides. On 13 November he wrote: 'Made sketches and wrote description of type justifier casting spaces as called for and inserting them in the line automatically.'

## 4  *About his illness*

The inventor treated early bouts of ill-health as minor set backs. After the serious attack of pleurisy in 1888 he resumed his punishing schedule and there were no further comments about his health until the summer of 1894. By Monday 30 July the symptoms had become so severe that he had to delegate the management of his factory. In his diary Mergenthaler wrote: 'Left for Blue Mountain on advice of Dr Thom Opie on account of what he called "lung trouble".' Mergenthaler returned to Baltimore on Friday 14 September, 'improved in general health but not in regard to the cough. Arrived at 159 Lanvale Street at about 12 am where Mrs Mergenthaler had moved during my absence. Find the house in excellent condition, everything looking like new.' Two January 1999 photographs show in figures 23 and 24 respectively, the property with a substantial coach house behind the house [street cars ran along the road to the side of the house in Mergenthaler's time] and a badly weathered aluminium memorial plaque by the front door.

Dr Wm Osler told Mergenthaler on 6 October that his upper right lung was affected and that his sputum contained considerable numbers of bacilli; the trouble was still local and he advised the inventor to go to Saranac Lake in the

Figure 23   The Mergenthaler House at 159 Lanvale Street West
*Source: Author's photograph*

Adirondack Mountains, in northern New York State and put himself into the hands of Dr Trudeau. He went to New York on 16 October for the annual meeting of the Mergenthaler Printing Company and from there went to Saranac Lake to consult Dr Trudeau.

On 26 November he put Carl Muehleisen in charge of the technical department at the factory and noted that Mr Greenleaf kindly consented to act as Treasurer. Then he left Baltimore and spent the winter at Saranac Lake with his family. They stayed at the Baker Cottage which had been occupied by Robert Louis Stevenson, the Scottish author, seven years earlier. His health fluctuated and it was too cold in the northern winter so he sought a warmer climate.

On 28 December 1894, before sailing for Germany to attend a hearing, Dodge wrote to Mergenthaler at Saranac Lake to confirm an order for 100 machines for which the inventor would be paid $875 each. There was no mention of or enquiry about his health.

No diary entries for 1895 seem to have survived, but on 18 December 1895 an article in the *Fourth Estate* reported that Mergenthaler: 'would be an ideally happy man if it were not that he suffers from very bad health due to his

Figure 24   Memorial plaque by the front door of the Mergenthaler House
*Source: Author's photograph*

tremendous labours and the carelessness of self, unfortunately characteristic of genius. The illness of the inventor does not keep him from working over his machine and it is said of him that when physicians insisted that he try a change of climate he said that he would rather die than be separated from his shop.' This indifference to his health was certain to end in disaster; on 9 January 1896, when Mergenthaler started to take Dr Shade's Chloridum cure, he was told that both lungs were diseased. The right one which was worst had a large cavity on the apex. His weight without coat was only 113 lb.

The treatment must have been unsuccessful because he went to New York to take Dr Edsom's cure on 14 March 1896. He returned from New York on 4 April in what he believed to be a slightly improved condition. However, Mergenthaler could not let go, particularly over his complaints about Dodge's management and his feeling that the company had cheated him out of royalties. Apparently he interrupted his treatment in April and May to attend business meetings in Washington and New York.

The inventor may have finally got the message that year. On 24 June he left home for Prescott, Arizona, arriving at 10 am on 30 June, 'in fairly good condition.' He wrote in his diary on 13 July that he was having a summerhouse built near the town with the idea of camping out; he spent six months there with only a guide for company. Naturally he did not rest while in Arizona; he started to write his autobiography. That he meant to write a contentious account is shown by the postscript to a letter to an unidentified acquaintance which stated: 'Am fairly well and working hard on that history. It will cause a

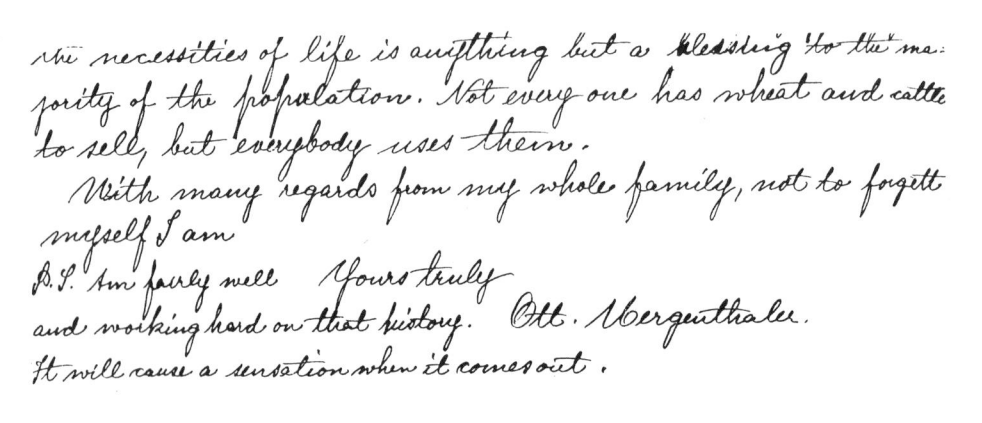

Figure 25  Fragment of a letter from Mergenthaler to an acquaintance
*Source: Mergenthaler family papers*

sensation when it comes out.' See figure 25. Soon after arriving he was approached by some young men who claimed to have made a great strike in a gold mine at Phoenix and wanted money for development. The gold taken out for assay showed well – his partner saw to that – but in operation the mine never came near paying and that hole in the ground cost him $11,000. On 20 August he wrote to Anna, his sister-in-law, that he was suffering less in Arizona than in Baltimore but that the cough was probably as bad as ever. Both Anna's family and his had suffered from excessive heat that summer and he commented that it was very hot in direct sunlight. He must have felt chilled all the time because he wrote: 'The weather was never very warm here, 88 degrees being the highest my thermometer recorded out here in the pavilion.'

In seeking a more suitable climate Mergenthaler moved (probably early in 1897) to Deming, New Mexico, where his family joined him. Figure 26 is copied from a badly aged photograph of the Deming house showing Ottmar and Emma on the porch and the children on the front steps. Mergenthaler continued to work on his autobiography with the help of Otto Schoenrich, his sons' tutor. By November 1897 it was nearing completion when a prairie fire destroyed the house and most of his papers, including the only copy of his manuscript; the family were lucky to escape with their lives. Mergenthaler made some diary entries on plain paper; 3 November 1897: 'Dwelling gets on fire and destroys completely the whole of the building and all its contents including stock certificates, drawings, patent cases etc, in short everything except Emma's jewellery and a few diaries.' However, the fire at Deming was not mentioned in the biography, but that could have been because Mergenthaler was so angry with Dodge that it drove other matters from his mind.

On 27 November he wrote: 'Started the meat and hot water cure known as the Salisbury treatment. Pulse 130. Condition weak.'

Despite the disastrous fire and his poor health Mergenthaler remained remarkably cheerful in his correspondence. Although doctors had told him that he was terminally ill he refused to give up hope and obviously thought that he would recover. On 4 December 1897 he wrote in old German script to a friend in Baltimore: 'As you can see I'm still alive and hope to stay alive for a long time. I also hope to shake my illness.' He continued in a light-hearted way: 'Christmas is near, the time we remember our friends. Please send me a pair of cuff-links for Christmas (gold, of course) costing about $5. You'd hardly recognise me if I came back to Baltimore. I don't drink beer or smoke cigars any more. Illness makes people moral.'

Figure 26  House at Deming, NM, destroyed by prairie fire in 1897
Source: *Mergenthaler family papers*

On 17 December he received the first diet sheet and instructions from Dr Winsor and noted: 'Pulse 100. Condition weak, stomach working fairly well. No particular effects noticeable to date except an almost continuous hunger, and a diminution of the expectoration, particularly in the morning. Have a keen appetite for meat.'

On Christmas Day he wrote: 'General condition like on previous note. Pulse 90. Condition weak, stomach fairly well but still generating gases and acids, particularly after the noon meal or rather one and a half to two hours thereafter,' and on New Year's Day 1898 noted that his condition was unaltered. He remained in Deming for a few more months to complete the 'Biography of Ottmar Mergenthaler and History of the Linotype, its invention and development,' with Otto Schoenrich. During that time his condition was little changed but by 10 March 1898 his weight had dropped to $100\frac{1}{2}$ lb. He thought that the apparent loss in weight could be due to wearing different clothing. There is no record of his stripped weight.

Mergenthaler finally left Deming for Baltimore on 16 April travelling via New Orleans. On 7 May he wrote that he did not seem to have suffered any bad effects from the change of climate and was in a fairly satisfactory condition. His stomach was working better than at any time since starting the Salisbury cure; he had no fever, sleep and appetite were good and his mind bright and collected but he was very weak, particularly in the legs.

It was not long before Mergenthaler was putting himself under a strain again. On Sunday 17 July he went to Ohio with Abner Greenleaf first to Painsville to see demonstrations of a basket making machine and then to Findlay to see how weldless tubes were made. They did not return until the following Saturday. The inventor apparently spent the next few months designing his logotype machine, which took until mid-November. Then he spent a hectic week in New York starting on Monday 28 November at the office of the Mergenthaler Linotype Company's factory with discussions about an interference proceeding. Mergenthaler probably expected this meeting to be acrimonious – he noted in his diary that Mr J. H. Watson acted as his attorney with Dodge and Rob Rogers acting for the Company. The rest of the week was spent on examining several typesetting machines in detail including the Empire, the Dow and the McMillan. The following Saturday he wrote in his diary: 'Returned to Balto much exhausted and resolved not to travel again unless in a materially improved state of health.'

This was the final diary entry in the family papers.

## 5 Recognition during his lifetime

There are several forms of recognition some of which carry no financial reward but are nevertheless deeply appreciated. This implicit form is seen in the attitude of Mergenthaler's workmen who respected him and admired his mechanical and creative skills. Encouragement from those in authority was equally important to Mergenthaler. The 'jeremiad of complaints' that he endured when he was working all hours to fill the early orders did not motivate him. After Reid's initial enthusiasm with his forecast of a castle on the Rhine the reality of a president who refused to pay his royalties and a company that induced him to accept a reduction left him embittered. His complaint: 'Repudiation over and over again!' was due as much to a lack of encouragement as from arguments over his royalties. Reid and Dodge both wanted to make Mergenthaler more compliant, though such a policy could well have backfired. On 4 July 1894, Clephane tried to reassure him when he wrote: 'Reputation, is a thing, though sought after by many is attained by only a few. By your transcendent abilities you have won it; now you must not by reason of jealousy of two or three, temporarily in authority, permit a lack of effort to in any way tarnish, or diminish it.'

The growing popularity of the Linotype after the mid-1890s was an endorsement with a price tag because he was receiving some $50,000 per year in royalties. This income led to the rumour that he was a millionaire, allowed him to pay off the $10,000 mortgage on his factory and to buy his $30,000 house at 159 West Lanvale Street. It also gave him the means to pay for the best medical treatment. Judgements in favour of the Linotype by Judges Coxe and Acheson in the cases against the Rogers Typograph were endorsements of his creativity.

Mergenthaler's brilliance was recognised by learned institutions during his lifetime, although they were not mentioned in the biography. In 1890 the Franklin Institute of Philadelphia awarded him the Elliott Cresson gold medal for the invention of the Blower Linotype and the following year the City of Philadelphia awarded him the John Scott medal for improvements to the Linotype. He also received a medal from the Cooper Union of New York. Images of both faces of the Cresson and Scott medals were printed on the cover of the Mergenthaler Company catalogue, see figure 27. The inventor was recognised and respected by experts throughout the world; the Linotype was awarded a gold medal in Antwerp in 1894 as shown by the British Linotype Company advertisement of figure 28.

Figure 27  Cover from the Catalogue of Ott. Mergenthaler and Co

48　　　　　*THE BRITISH PRINTER ADVERTISEMENTS.*

## THE Linotype Composing Machine.

GOLD MEDAL—ANTWERP, 1894.

### London Daily Papers.

The **Daily Telegraph** has Twenty New Improved Quick changing Linotype Composing Machines now running.

The **Globe** has been entirely set by Linotype for 18 months.

The **Financial Times** has used Linotype Machines for a year.

The **Morning** has just put down Linotype Machines.

### Provincial Daily Papers.

The following are all set up by the
**LINOTYPE:**

Aberdeen Evening Gazette.
Aberdeen Free Press.
Belfast Evening Telegraph.
Belfast News Letter.
Birmingham Daily Gazette.
Bolton Evening News.
Cambria Daily Leader.
Cardiff Evening Express.
Darlington North Star.
Eastern Morning News.
Hull Daily News.
Irish Daily Independent.
Leeds Mercury.
Leicester Daily Mercury.
Leicester Daily Post.
Liverpool Mercury.
Manchester Courier.
Manchester Evening Mail.
Manchester Evening News.
Manchester Guardian.
Newcastle Daily Chronicle.
Newcastle Daily Leader.
Newcastle Evening Chronicle.
Northampton Daily Chronicle.
Northampton Daily Reporter.
Nottingham Daily Guardian.
Nottingham Daily Express.
Nottingham Evening News.
Nottingham Evening Post.
Oldham Evening Chronicle.
Scarborough Daily Post.
Scarborough Evening News.
Sheffield Daily Telegraph.
Sheffield Evening Telegraph.
South Wales Daily Star.
Sporting Chronicle (Manchester).
Sunderland Daily Herald.
Western Daily Mercury.
Western Mail (Cardiff).

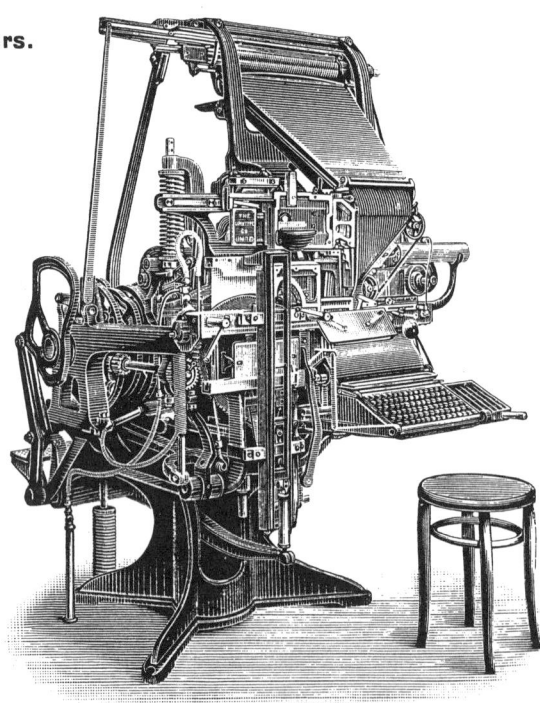

The Linotype Co., Ltd.,　6 Serjeants' Inn, London, E.C.

Figure 28　Linotype Company advertisement with Antwerp Gold medal
*Source: The British Printer, March/April 1895, Advertisements, p 48*

When Mergenthaler returned to Baltimore from New Mexico he gave the order for printing the shorter version of his biography to a local printer, the Friedenwald Company, the first book printer in the world to install a Linotype. One thousand copies of the 72-page pamphlet were printed at the inventor's expense and distributed among his friends. This publication quoted correspondence with Reid when he was head of the Mergenthaler Company and in effect blamed persecution by Reid with so affecting his health that he had to give up work.

An immediate reaction to the biography came from E. V. Murphy, one of the original members of the Washington group and sometime director of the Mergenthaler Printing Company. On 29 September 1898 he wrote to thank the inventor for his copy. He accepted the truth and force of the account and was much gratified at the credit given to mutual friends such as Clephane, Hine, Greenleaf and others. Murphy, who thought that the inventor had in many cases been outrageously and infamously treated, had always tried to speak up for him; he recognised his marvellous ability as an inventor and his great capacity as a man of affairs and if the Company had followed his advice thousands upon thousands of dollars would have been saved and it really would have been on a practical commercial basis much sooner than it was. All must now recognise that the determination to put out machines prematurely and against advice was a very costly folly. He hoped that full justice may yet be done, that Mergenthaler would soon be restored to perfect health and, after further thanks, signed himself: Your friend and admirer, E. V. Murphy.

At about this time Reid paid Mergenthaler a back-handed compliment but the inventor may not have lived to appreciate the irony of the situation. Reid was a public figure who was often sent abroad as a special delegate and later became United States Ambassador to England. He considered the book to be so damaging to his reputation that he engaged agents to acquire and destroy every copy. On 17 December 1974, The *Baltimore Sun* printed a letter from Hugo Dalsheimer, son of Simon Dalsheimer, President of the Friedenwald Company in Mergenthaler's time, who recalled:

> One of the older men in the office of the Lord Baltimore Press [formerly the Friedenwald Company], who knew Mergenthaler, had a copy. He told me that he was offered and refused to take, $75 – a tremendous price at that time for the thin paper-backed pamphlet. Subsequently a "friend" borrowed his copy. He never got it back.

## 6  *His death, his will and estate*

Mergenthaler continued to decline during 1899 but remained alert, took a keen interest in his business and current affairs and did not give up work completely until shortly before he died. On 16 October 1899 the *Baltimore Sun* announced that he was seriously ill. He died twelve days later, at about 9am, on Saturday 28 October, at the family home in Baltimore with his family at his bedside.

Clephane was distraught when he heard about Mergenthaler's death and wrote to Emma from his Washington address on 30 October 1999:

> My dear Mrs Mergenthaler
> Sad indeed were the words of the telegram from Mr Greenleaf announcing '*our friend* is dead.' So kind and thoughtful in Mr Greenleaf to use these words – 'our friend.' For over twenty years your beloved husband, Mr Greenleaf and myself have been most intimately associated, and never, during the whole of that time do I recall that we ever had the slightest difference.
>
> From the time I first met your husband, quick to recognise his great genius, continued association afterwards had the effect not only of still further impressing upon me that genius, but of exhibiting those grand traits of character which he possessed to such a remarkable degree – sterling honesty, intense love of justice and right, and imperishable friendship for those he felt worthy of his esteem.
>
> The World will, of course, not be tardy in giving recognition to his great genius, but only his friends, those who knew him intimately, can do justice to his great worth as a man, a husband, father and friend.
>
> How sad the thought that one so talented, and who was always so willing to give the world the benefit of his great gifts, should, at such a comparatively early age, be taken from the scene of action. The ways of Providence are indeed inscrutable!
>
> It is a great comfort, however, to know that he lived long enough to have knowledge of the fact that the world appreciated the great work he had accomplished, and had during his lifetime placed his name high on the roll of enduring fame.
>
> You will please accept my sincere sympathies in your great bereavement.
> Yours very sincerely,
> Jas O. Clephane

The funeral started from the house at 10am on Tuesday 31 October 1899. The service was conducted by the ministers of the Trinity German Lutheran Church and the Brown Memorial Presbyterian Church. Active pall-bearers were selected from the factory of the Ottmar Mergenthaler Company. Among the honorary pall-bearers were four of the inventor's closest associates from the Washington group; James O. Clephane, Abner Greenleaf, Lemon G. Hine and Frederick Warburton. Mergenthaler was buried at the Loudon Park Cemetery in Baltimore where other members of his family are also buried. There is no evidence that either Whitelaw Reid or Philip Tell Dodge was present, but in the circumstances that was hardly surprising.

Mergenthaler's will consisted of three parts. First, he gave $2,000 free of tax to the German Orphan Asylum. Second, he bequeathed one third of the residue of his estate to his beloved wife Emma, but realising her lack of experience in money matters he urged her to seek financial advice from Abner Greenleaf. He warned her against loaning money to relatives or friends, against endorsing notes or other obligations, and against giving any part of or interest in her property to anyone whatever, except in possible contingencies to her children. Third, the remaining two thirds of the residue was put in trust with the Safe Deposit and Trust Company of Baltimore and Abner Greenleaf, 'to be applied to the liberal support maintenance and education of his children.' As each son attained the age of twenty-one years he was to receive his share of the principal. However, the trustees would continue to hold the share of each daughter for the term of her natural life and, when she reached the age of twenty-one, would pay the net income from her share into her own hands; the trust on her share to cease on her death and bequeath the same to her children then living. Any income not expended by the Trustees during the minority of his children to be added to the principal.

Mergenthaler's estate came to just under $500,000 at his death but the family received an estimated $50,000 per year in royalties, probably until the early 1920s and his Baltimore works continued in operation for several years. This meant that Emma received about $160,000 from the estate with an annuity of some $17,000 before taxes. Long after her children had left home Emma married William A. Stauf, a retired army officer whom she supported. Apparently he helped her to run through her fortune quite quickly; Greenleaf had died years before she remarried. By this time the elder sons had died and she was so short of money that her two younger children each gave her $3,000 a year to help out. Thus Ottmar's fears were well founded and it was probably to protect his daughter against a similar fate that he entailed her portion. A wise decision because Pauline divorced Rody Perkins and resumed her maiden name.

## 7   *Posthumous recognition*

In his letter to Emma after Mergenthaler died, Clephane had forecast that: 'The World will, of course, not be tardy in giving recognition to his great genius.' Unfortunately he was wrong, possibly because of the antipathy between Mergenthaler and Dodge. The Company did comparatively little to honour him until 50 years after the first machine was installed at the *New York Tribune*. In general it has been people outside the company who have campaigned for recognition of his achievements, particularly in Hachtel and Baltimore. The obituary in the *Inland Printer* of December 1899, claimed that his name would live for all time and that the Linotype would live forever. Since the computer displaced hot metal Mergenthaler's name is mainly known to people with an active interest in printing history, but at least his genius is still widely acknowledged in Germany.

Emma was interviewed by a Canadian newspaper in 1906 when she spent a holiday with her two younger children at the Thousand Islands. The article which carried the headline: 'How **Otto** Mergenthaler invented the Linotype' noted that the Linotype had been pronounced 'the most remarkable machine of the century' by the *London Engineering News*.

George Iles included a 40-page illustrated biography of Mergenthaler in his November 1912 book, *Leading American Inventors*. This includes anecdotes about the inventor that were repeated in articles about him on the 50th anniversary of the installation of the first machine at the *New York Tribune*.

On 11 May 1917, the sixty-third anniversary of the inventor's birth, Emma accompanied by her daughter Pauline (Mrs Rody Perkins) went to the Shepard public school in Chicago where Pauline unveiled a bronze bust of her father by Hans Schuler, a Baltimore sculptor who had been an intimate friend of Mergenthaler, see figure 29. Pauline spoke feelingly of her appreciation of the honour to her father's memory. It was suggested that a replica of the bust, which was based on the 1894 photograph of the inventor, should be placed in the Smithsonian Institution in Washington where the original Linotype had recently been installed. Later that day Emma was elected President of the Mergenthaler Association.

Figure 29   Emma and Pauline at a ceremony honouring Mergenthaler
*Source: The Fourth Estate, 19 May 1917, p 14*

A petition was presented to the Board of School Commissioners of Baltimore on 10 June 1920, to establish a central school of printing to provide courses to replace the old apprentice system that was rapidly disappearing. The petition concluded: 'The greatest invention in the typographic art since the days of Gutenberg was made here in Baltimore, the typesetting machine by Ottmar Mergenthaler. Nothing has been done to commemorate him and we suggest that the school, if established, be called "The Mergenthaler School of Printing".' Simon Dalsheimer was one of the members of the trade in Baltimore who

Figure 30  The refurbished school house at Hachtel (c 1925)
*Source: Linotype Company publication*

actively supported the project. The Ottmar Mergenthaler School of Printing opened in September 1923 across the street from the jail. Since being built in 1872, the small two-storey red brick building had housed several different schools. The school of printing was originally intended to give day and evening printing courses to boys and men of the city, but was opened to girls during the second world war.

A meeting was organised in Hachtel on 9 November 1924 to mark the 25th anniversary of Mergenthaler's death. The schoolhouse where he was born was dedicated as a Mergenthaler museum, as shown by figure 30, and the German Linotype Company donated a memorial plaque.

In 1936 several printing journals carried reminiscences of the first Linotype, and its inventor, 50 years after it was installed at the *Tribune*. In December 1936 *The Linotype News* reported that on 23 November, the hundredth anniversary of the institution of the American patent system, Ottmar Mergenthaler had been named alongside Alexander Graham Bell, Thomas Alva Edison, Samuel Morse, Wilbur Wright and others as one of the twelve greatest inventors of America.

Eugene George Mergenthaler, who died in 1919 at the age of 33 years, left part of his estate to finance a memorial to his father. Twenty-three years later his bequest helped to build Mergenthaler Hall, see figure 31, a four-storey biology laboratory at Johns Hopkins University in Baltimore, completed in 1942 at a cost of $400,000. Herman and Pauline presented the bronze memorial plaque shown in figure 32; it was unveiled in 1950.

The largest monument to the inventor in his adopted city of Baltimore is The Mergenthaler Vocational Technical High School (Mervo-Tech), also known as the Mergol, which occupies a 17-acre site at Hillen Road and 35th Street to the north east of the city. The entrance to this enormous building is shown in figure 35. This major educational facility, which opened in September 1953, was formed by merging three vocational high schools, including the Ottmar Mergenthaler Printing School. Its catchment of some 2,000 students is drawn from all sections of the city and all ethnic groups. A $20 million refurbishment started in 1998. Unfortunately too few residents know the name Mergenthaler or appreciate its importance. [In 1992, when a clerk at the Hall of Records was asked for a copy of Mergenthaler's will he said, 'Did you know that there is a school of that name here in Baltimore; is there a connection?' He was amazed to hear that Mergenthaler possibly made a greater contribution to civilisation than any other resident of the city.]

Figure 31   Mergenthaler Hall at The Johns Hopkins University
*Source: Copied from a photograph taken by Professor Corban Goble*

In May 1954 the centenary of the inventor's birth was commemorated on both sides of the Atlantic. In England on 11 May, *The Times* reported: 'Readers of newspapers, periodicals and books throughout the world have reason to remember with gratitude Ottmar Mergenthaler, who was born 100 years ago. Mergenthaler's genius gave the printing industry a completely mechanical means of setting type that made possible one of the great strides forward in its long history.' A festival book in English and German versions was commissioned by the Linotype companies. The Hachtel schoolhouse was refurbished with a timber-clad exterior and rededicated in a ceremony sponsored by the German Linotype Company and attended by Herman, Pauline and her daughter, compare figures 1, 30 and 33. The German Post Office issued the 10 pfennig

Figure 32   Memorial plaque to Ottmar Mergenthaler
*Source: The Ferdinand Hamburger, Jr Archives, The Johns Hopkins University, Baltimore, Maryland*

Figure 33    The refurbished school house at Hachtel (c 1954)
*Source: Linotype Company publication*

Mergenthaler stamp, shown in figure 34 and set up a post office for franking stamps in Hachtel, a village too small for a regular post office. A new type face named *Mergenthaler Antiqua* was created in Germany.

In 1982 Mergenthaler was recognised by the Patent and Trademark Office, Washington, DC which installed him in the National Inventors Hall of Fame for his first machine that used circulating matrices. It recognised the Linotype as 'one of the most significant technological advances of the 19th century.' By contrast, Schlesinger noted that in 1920, the American Newspaper Publishers Association unanimously voted to propose Ottmar Mergenthaler as a candidate for the Hall of Fame [of Great Americans], but he was not elected.

The United States Postal Service issued a 32 cent Mergenthaler stamp on 22 February 1996 which, unlike the German 10 pfennig stamp shown at the top left of figure 34, did not represent a particular year. It was one of four stamps issued to honour American inventors who were Pioneers in Communications, top right group of figure 34. This issue was not a sudden decision. Several dedicated men including Carl Schlesinger and Corban Goble submitted proposals for a Mergenthaler stamp. In fact Schlesinger's persistent campaigning so affected the Postal Service's Stamp Advisory Committee that in 1994, with their usual reply of 'under consideration', the envelope was addressed to 'Carl Schlesinger, Ottmar Mergenthaler Fan Club'. Once the concept had been accepted there were problems with the actual illustration. The detail at the bottom right of the draft design, showing the prototype of the Blower machine, was turned on its side. Schlesinger immediately tried to get the artwork changed, but was too late to stop the advance publicity; the official US Stamp catalogue still showed the machine on its side. The artist, Fred Otnes substituted an illustration of the second band machine [the first line casting machine] which was a suitable shape to fill the available space. It is a pity that these stamps are almost completely black which obscures much of the detail.

The inventor was featured in two radio broadcasts in the mid-1930s and in the 1952 film Park Row but those presentations contained so many distortions of fact that they are discussed in Chapter 9.

## 8  *Mergenthaler – the man*

This account of Mergenthaler and his machine has shown that he was a complex character. Sensitive, yet brash; cautious, yet headstrong; honest and tactless; he was one of the great inventors of the 19th century and possibly his own worst

Figure 34  Mergenthaler postage stamps - top left: Germany 1954;
top right: USA set 1996; bottom: Mergenthaler first day cover

enemy. It would be trite to concentrate on his many virtues and ignore his faults because without this mixture of positive and negative traits he would have been less than complete. He succeeded in spite of his faults and possibly because of them. Had he been less stubborn he might have spent a long and healthy life as a watch and clock maker in a German village.

Mergenthaler was a perfectionist with many virtues including honesty, loyalty and ambition. An aspect of his honesty was the stubborn streak that did not allow him to compromise if he felt sure of his position. He showed both honesty and ambition when he stood up to his father about his future. Having won the concession of becoming an apprentice in a craft that he found acceptable he strove to become more expert than his fellows, took optional extra evening and weekend courses and was rewarded with a journeyman's pay before he finished his time. He was friendly with the other apprentices and seems to have been on good terms with everyone.

Mergenthaler has unfairly been called a draft dodger because he left Germany before he could be conscripted into the Prussian army. At the time he was just one of many who went to the USA to seek their fortunes and were against serving in an army dedicated to imperialism rather than defence. He travelled steerage, the cheapest way, and worked his passage by repaying his fare out of his wages. When he started working for August Hahl he had no experience of electrical instruments but soon became so expert that he was made foreman before he was 21 years old. He was lucky to have a place to go in America but was promoted on merit. He also made models for inventors who were applying for patents. This work inspired him and formed the basis of his future.

Although speaking only German when he came to America he had learned to speak English fluently within a few months but he did not always appreciate the nuances of the language. He used English in his business correspondence and in his diary but never mastered English spelling. He showed his loyalty to his new country by becoming a citizen of the United States in 1878. From that time he considered himself to be American but still liked to spend his leisure with German friends. In the early days he was often short of money and probably thought it character forming to have to make one's own way in the world; just as he had had to do. It is in his attitude to money that Mergenthaler threw out conflicting signals. He could be generous at one minute and rather mean the next. Apparently he sent money back to Germany to pay for studies by his elder brother Adolf and he brought Fritz to America, but he seemed far from generous

Figure 35   Entrance to Mergenthaler Vocational Technical High School
Source: *Author's photograph*

when his half-brother wanted money rather than shares to take charge of the Locust Point works. The mutual stubbornness of the brothers ensured that the breach was never healed. He paid Muehleisen much less to run his works than he had been paid to run an experimental shop, even though Muehleisen was an inventor in his own right. Nevertheless his men remained fiercely loyal to him; in particular, Ernest Girod and Ferdinand J. Wich (both of whom erected machines in Europe) sent him reports about the machines at the *Tribune* office which helped him to make them more rugged.

The inventor was very grateful to Clephane and gave him a valuable present and cancelled his debts, but in his will he gave a small bequest to an orphanage and left nothing to anyone outside his immediate family.

Mergenthaler obviously enjoyed the affluence that came with success and adopted a lavish life style. He bought a large house in a fashionable area and drove a rig with a spirited pair of horses. In 1895 the *Fourth Estate* wrote that he was reputed to be a millionaire and in a speech to Linotype salesmen in 1913, long after the inventor's will was published, Warburton falsely claimed that Mergenthaler had died a millionaire.

Mergenthaler showed a lack of gratitude to Hahl for his help in 1872 and his complaint that Hahl received the same share as he in the National Machine Printing Company was rather spiteful being based on the fact that he had several inventions to his credit and Hahl had not.

Despite bad health he pushed himself to the limit and never stopped trying to invent and to learn. He was friendly with printers in Baltimore, particularly Charles W. Schneidereith and Simon Dalsheimer of the Friedenwald Company, who told him about requirements in the trade. On 6 October 1896 during his final illness he went to Washington to submit an informal offer for the German Linotype patents. He discussed the matter with Devine, Clephane and Hutchins. This is important because it was reported that the inventor and Hutchins had differences, but there does not seem to be any written record of such disagreement.

Finally, there was the question of Mergenthaler's bitterness towards Reid and Dodge which although justified probably hurt him more than it hurt them. Despite his skill as a craftsman and his genius as an inventor Mergenthaler lacked higher education and because his spelling was often faulty they probably treated him as ignorant rather than unlettered. With his imperfect understanding of English idiom he tended to give offence unintentionally, as when he did not take Reid's 'earnest advice', a facet of his ability to remember praise and ignore criticism. For his part, Reid's response was malicious and typical of a rich man using his financial muscle to attack a less affluent adversary. In that case only Mergenthaler's stubbornness saved him – had he been compliant Reid would have tricked him out of everything. His complaints about Dodge were of a somewhat different character; although Dodge withheld his royalties for months he was more secure because by then he had an established workshop. Dodge wrote to Herman on 23 September 1920 about his father: 'It was always my sincere desire to do justice to him. I am sure he was misled, in matters which I need not detail, by outside advice with resulting friction and with loss of possible profit to him.' Regardless of this mild statement his ill will obviously carried on after the inventor's death because the company made no attempt to honour Mergenthaler during Dodge's term of office.

# SEVEN

## *An overview of the British Linotype Company*

The foundation of the British company, which was independent of the American companies, was principally due to Stilson Hutchins and Joseph Lawrence. The launches of the Linotype in the USA and the UK were markedly different. The American syndicate did not publicise the Linotype until they decided to sell the manufacturing rights to an overseas buyer. The British syndicate, on the other hand, believed that they were buying a perfect machine and embarked on a saturation advertising campaign to invite investment and generate sales. Their publicity included American claims that were a mixture of over optimism and downright lies. Probably the most blatant were claims about 'the present level of perfection', when the Americans knew that they were dumping obsolescent machinery on to an unsuspecting customer. However, Hutchins retained his credibility; ten years later, on 14 July 1899, he was an honoured guest at the formal opening of the new Linotype plant at Altrincham, Cheshire.

### 1   Background to the British launch

There was little news about Mergenthaler's machine from 1885, when the syndicate took over management of the National Typographic Company, until the machine had been in regular use at the *New York Tribune* for nearly three years. The short reports in American trade papers tended to be far from specific although some did ultimately appear in contemporary British journals. Even experts in the British printing industry, as well as the general public, were largely unaware of the Linotype before it was demonstrated in the United Kingdom in the early summer of 1889.

Lawrence was the first Englishman to show an active interest in the machine. He returned to London in 1887, after spending two years in the USA, and started two railway trade papers. Operating costs were so high that he returned to America, probably during late summer 1888, to see if Linotype setting would be more economical. It is not known how he heard about the machine but he may have met Bullen and seen the slugs that he was taking back to Australia. Lawrence probably wanted a few Linotypes for his own use and might have considered starting an agency in Britain.

Hutchins must have realised that Lawrence's interest in the Linotype could be developed into a business opportunity that would dwarf the profit on the sale of one or two machines and help to restore American finances. His brilliant scheme did not include starting a British branch as assumed by Mergenthaler. Instead, he sold the manufacturing rights to a syndicate which founded the Linotype Company in the UK took all the risks and paid handsomely for the privilege. He also offered 60 Blower Linotypes to the British, so helping to dispose of the surplus machines that were about to be superseded by the 'improved' Linotype. Although Lawrence was the moving force behind the negotiations, the British syndicate was led by his friend Jacob Bright, MP (see biographical notes).

Before approaching the Board, Hutchins must have reached agreement in principle with British interests, and, having obtained their approval, he would have had to draw up detailed plans to handle the practical and legal aspects of the deal. The negotiations called for a delicate balancing act combining precise timing with absolute secrecy. Contemporary reports indicated that there were at least four machines in the United Kingdom when the Linotype demonstrations began in early summer 1889: two in London, with other sites in Manchester and Glasgow, but there was no information about when or how the machines were imported and installed. Hutchins's operating plan must have included the following factors, though not necessarily in the order given:

1 Negotiate the sale of manufacturing rights in the Linotype, outside the Americas, to a British Linotype syndicate,

2 Offer 60 Blower Linotypes at $1,000 each (£200 sterling) claimed to be cost price but actually much less than the cost of production, see 'currency' in glossary,

3 Send British engineers to the USA to be trained on the Blower Linotype. Matthew Whittaker (see biographical notes) was reputedly taught by Mergenthaler, and was said to have erected the first machine in the United Kingdom in 1889,

4 Install two Blower Linotypes, numbers 171 and 183 (according to Whittaker) at 37 Southampton Buildings, Chancery Lane and set up an office with son Lee, an attorney-at-law, at those premises. Bring expert Linotype operators and engineers from the USA to produce the railway papers and give early demonstrations for newspapermen and printers,
5 Assign, set up and run other demonstration sites in the UK,
6 Insert complimentary semi-technical articles in the American press before the start of the British advertising campaign, including previously cited articles in the *Scientific American* and the *New York Tribune*, which were really covert advertisements for the Linotype,
7 Demonstrate the Linotype to a team of United Kingdom experts in London, take them to see the machines working at the *New York Tribune*, and have them cable a favourable report,
8 Quote favourable comments about the Linotype in a country-wide newspaper advertising campaign, before issuing the company prospectus.

Hutchins was reported to have sailed for Europe in October 1888, but there is some doubt about the actual date. On 20 July 1889, the *Railway Herald* claimed: 'considerable portions of this journal and the journal with which it is affiliated have been set in type through the instrumentality of the Linotype.'

## 2  *Preparing to launch the Linotype Company*

On 20 June 1889, the National Typographic Company assigned the patent rights in the Linotype Printing machine in all countries of the world except North and South America to Jacob Bright. This assignment, notarised on 24 June 1889, gave the British syndicate authority to deal in Linotypes.

Before the end of June 1889, probably just after the above assignment was notarised, the Linotype syndicate opened a new showroom at 52 New Broad Street in the City of London and took a machine from Southampton Buildings to the new site. Demonstrations were held at both installations. Employees at Southampton Buildings complained that losing one of the machines interfered with production of the railway magazines.

A typescript in the papers of John Southward (see biographical notes) gave the experiences of John Dawson who was invited to edit Lawrence's railway magazines. He found only one Linotype at the site and the only skilled operator was Machen, a smart young American, about twenty years old; a noted 'swift' (known as a 'whip' in the UK). He was initially paid £10 per week, at least three times the average wage of a top London compositor. Later he was paid £8 per week and guaranteed his return fare to the United States. Other American operators included Bailey (a teenager assigned to Paris), Beekman (from the *New York Tribune*) and Harry G. Leland, who had worked for Mergenthaler in Baltimore. Compositors were paid two guineas per week [a little more than $10] to learn how to operate the Linotype. The company also hired a young man called Grant (not a compositor) on the same terms, but he crippled a machine at the outset and was fired. A promising trainee listed the shortcomings of the Linotype and noted that the editor resigned because of unsatisfactory output – he had nightmares, with Linotype machines sitting on his chest.

Publicity started quietly, probably in June 1889, with demonstrations of the two Linotypes at 37 Southampton Buildings, because by the end of June, there were reports of the machine in periodicals and the daily press. These varied from short general comments in daily newspapers to quite long technical articles in trade and engineering journals. There were no pictures in the daily papers, but several periodicals printed the woodcut of a Blower Linotype operated by a young man, see figure 36. The unnamed operator might have been Machen.

The Linotype was demonstrated to William Gladstone, Grand Old Man of the Liberal party and three times Prime Minister, at 52 New Broad Street on 26 June 1889. He set up a slug with his name, made a short speech of welcome for the machine and said that he was 'staggered'. Inevitably, some newspapers claimed that he saw the machine at 37 Southampton Buildings. As leader of the Linotype Syndicate and a parliamentary colleague Jacob Bright probably invited Gladstone to the demonstration; both men were reported as being present. The demonstration to Mr Gladstone was widely reported, first as a news item in the popular press and later in every Linotype advertisement.

The syndicate engaged Mr John Charles Cottam, a company promoter who led a group of financiers, dubbed 'Cottam's gang' by the *Star*, to float The Linotype Company. Cottam and his partner Mr Ernest Orger Lambert had promoted several dubious companies. Their rôle became known when the press attacked the launch of the Linotype Company a few days before the first advertisements appeared. On 7 July 1889, the *Sun* (a new Sunday paper first published on 21 April 1889) reported in sneering terms, that Mr John Charles Cottam was to 'engineer' the promotion of the Linotype. Adverse comments even came from the USA. In November 1889, the *Printers' Register* printed the

*American Bookmaker's* comment that: 'the greatest promoter of new schemes in England was at the head of the enterprise; several directors withdrew after learning what was going on in the way of manipulating the shares.' The *Sun* continued its attack with the statement: 'The Linotype is to be in the composing room, what it was fondly hoped the Moldacott [a new sewing machine] would prove in the nursery – a combination of amusement for the young and gratification for their elders. The only fault of the Linotype is that it won't do

Figure 36  A woodcut of the Blower Linotype in the UK
*Source: Railway Press, 21 June 1889, p 12*

varied work. But that is a trifle – if Mr Cottam runs it.' In particular it was reported that Lambert: 'had already started on the warpath, with the firm intention of clearing his character of the imputation of belonging to the Cottam gang.' Despite this report of a split the *Sun* assumed that Cottam and Lambert would work together on the Linotype Company promotion. Cottam did not deny the assertion.

On 9 July 1889 the *Hawk*, a weekly magazine that concentrated on the sensational exposure of dubious enterprises, noted that a company would shortly be started to take over the patents of the Linotype.

## 3  *Advertising the Linotype Company in the British Press*

Newspapers and magazines were the only forms of mass communication in the late nineteenth century and hence the most direct means of targeting potential investors. Further, most newspapers were set by hand and often consisted of only four, six or eight pages. In this size of paper a full-page display advertisement, particularly if placed on the back page, could hardly escape the reader's notice, or that of anyone else in the vicinity.

Although the advertising campaign and legal procedures overlapped, these aspects have been separated as much as possible in order to simplify this account. The Linotype Syndicate delayed national press advertising until mid-July 1889 by which time significant articles had been printed in several magazines. This was saturation coverage; Southward complained that one could not open a newspaper at the time without seeing a Linotype advertisement. The £11,000 cost of this press campaign was confirmed in court by the advertising contractor who handled the publicity.

There were two phases to the campaign. The first set of advertisements headed: 'The "Linotype" Composing Machine', described the capabilities of the machine and stated that The Linotype Syndicate was now prepared to receive orders for use of their Composing Machines on royalty, without directly specifying the amount of royalty. The second set of advertisements headed: 'The "Linotype" Company Limited', presented the Company Prospectus and later the Abridged Prospectus, which both called for investment. None of these advertisements was placed in printing trade journals. *The Times* and other long established papers maintained a sober single column style. Some of the more recently founded journals adopted a more flamboyant format, but all newspapers carried the same general message.

The advertisements in *The Financial Times*, a 17½ × 24 inch broadsheet established in 1888, were probably more brash and comprehensive than in most other papers; they would have cost about £600 at contemporary rates, a substantial portion of the £11,000 spent on publicity. This London daily penny paper, which was changing printers at the time, did not appear on August 5 and 6. Except for the notices withdrawing the prospectus, the advertisements in this paper all occupied a full back page. Under 6-column banner headlines in capitals half an inch high; they appeared on:

| | | |
|---|---|---|
| Saturday, July 13 | information | Back page 6 |
| Monday, July 15 | information | Back page 4 |
| Saturday, July 20 | information | Back page 8 |
| Saturday, July 27 | information | Back page 6 |
| Monday, July 29 | information | Back page 6 |
| Tuesday, July 30 | prospectus | Back page 6 |
| Wednesday, July 31 | prospectus | Back page 6 |
| Thursday, August 1 | abridged prospectus | Back page 6 |
| Friday, August 2 | abridged prospectus | Back page 6 |
| Saturday, August 10 | prospectus withdrawn | Back page 4, top half |
| Monday, August 12 | prospectus withdrawn | Back page 4, top half |

The first *Financial Times* advertisement of Saturday 13 July 1889, much reduced in figure 37, used 6-column headlines to proclaim: A Complete Revolution in Printing; The "Linotype" Composing Machine; Now used in the Offices of some of the Principal Newspapers in America; Cost £400,000 to bring to its Present State of Perfection. Saves One Newspaper alone over £16,000 a year.

A set of general and specific claims followed, apparently directed at investors rather than printers. The claim that the Linotype raised the unit of typesetting from a letter to a bar or line of type, hence its name, was valid, but it did not supersede: 'The present system of setting up by hand.' The Linotype could easily outperform the hand compositor when setting straight text in one font, but lacked the flexibility to set complex copy.

Sixteen 'advantages' listed in the publicity overlooked some of the most valuable features of the Linotype and tended to emphasise trivia. Comparing the Linotype with the inventions of Gutenberg and Caxton was flawed because Caxton was not an inventor. It claimed that the rate of composition was raised at least sixfold, but then exaggerated the figure to assert that experienced

Figure 37  A Linotype Syndicate advertisement
*Source: The Financial Times, 13 July 1889, p 6*

operators had set matter up to ten times as fast as experienced compositors. A competent operator could set corrected matter all day long, five or six times as fast as hand compositors, but interruptions for recharging the metal pot, emptying the galley, collecting new copy and resetting the machine after a stoppage would keep the rate down.

Three items dealing with the costs of machine composition were not substantiated. The statements that the system would save at least 70% in cost and that everyone could afford the royalty, gave no details and did not specify the amount. The potential loss on displaced equipment was ignored. Further it was stated that less space was required, which meant a reduction in rent; only true if one could rent less space. In fact rents could rise if the building had to be strengthened to support heavy machinery. It was reported that operators in America were earning an average of 17% more in wages than hand compositors; but that was not necessarily a guide to wage rates in the United Kingdom.

The assertion that little skill was required to work the machine and that its adoption was inevitable was fatuous, particularly in view of the young man Grant who crippled a machine. A good Linotype operator had to have mechanical skill and a flair for print. The adoption of the Linotype was assured because the trade was ready for hot metal. Skilled operators would consider the ability to read a line in the assembler as a minor advantage and would not disturb their rhythm by breaking off to read lines.

Saving the costs of keeping large stocks of popular sizes of type was significant but claims about the supply and use of any size of type were not valid in 1889. The early machines were made for only one size of type. To change the font was a major task that called for replacing the whole set of magazine tubes as well as the matrices. To say that it dispensed largely with the need for stereotyping ignored the fact that stereotyping was used to make cylindrical moulds for printing newspapers on rotary presses; it emphasised the lack of know-how of the people drafting the publicity.

The statement that because the type metal was reusable there was little waste ignored the cost of reprocessing and rejuvenating that metal; dirty metal would clog the metal pot. The ability to cast line after line of the same matter without resetting was useful for composing page headings for books, but not necessarily for repeating complicated advertisements. It could take as long to separate and collate duplicated lines as it would to set each insertion separately. Comments about 'pieing' movable type were very important. Dropping a made-up page of movable type on its journey from the composing room to the press was a real disaster. Speed was most important to newspaper production, particularly when a story broke just as the paper was going to press.

An advertising campaign directed at printers would have emphasised: speed of operation; lower operating costs; the line setting concept and ease of locking up with the advantage that thin sorts such as i, l and t, could not break off the end of a Linotype slug; no shortages of type and every job cast in new type; automatic distribution of matrices; automatic justification leading to consistent spacing when newspaper articles were broken down into several short takes.

After the banner headlines and preliminary claims the page was laid out in two 3-column spreads consisting of complimentary quotations from many publications; fifty-one British, five American and one Australian.

The first piece, from the *New York Tribune*, noted that Whitelaw Reid had become American Minister in Paris and gave American prices in terms of Sterling: a royalty of £80 per machine per annum with a non-returnable deposit of £200 per machine. [At that time a three-bedroom cottage in an outer London suburb cost about £50 freehold.] It was claimed that the Typographical Trades Unions in America were fairly reconciled to the machine. (Presumably to challenge the London Society of Compositors and the Typographical Association.) It also stated that the Linotype had taken at least nineteen years to develop; this was from when Clephane started – not when Mergenthaler became involved.

There was yet another report about Mr Gladstone and the Linotype. The rest of the page was filled with short quotations from other journals. The *Morning Post* of 5 June stated: 'Its merits are forcibly attested by the practical demonstrations given under the auspices of Mr Hutchins at Chancery Lane.' The full report confirmed that Stilson Hutchins organised the demonstrations at Southampton Buildings.

The second and third advertisements on Monday 15 July and Saturday 20 July respectively, gave the same information but presented differently on each day. There was a new layout on Saturday 27 July. Information that had previously appeared in the top half of the page was compressed into the top third of the back page of a six-page paper. The rest of the page, under a full width heading, announced 'Extracts from Reports by Experts in the Printing Trade and Others', and once again gave the opinion of Mr Gladstone. On 3 July, the deputation which went to the USA cabled from New York: '... The machine meets with our entire approbation ... Wire if you want fuller information; all are delighted with machine.'

Below this, half the page was laid out in six columns of short articles in small-ad format, each headed: The Linotype Composing Machine (in capitals). The first, from the English Delegation, dated 21 June, was about their examination of the machines at Southampton Buildings, before they left for New York. Then Edward Wyman, Manager of the Hansard Publishing Union Limited, praised the machine, especially for the ability to read lines in the assembler and the certainty of accurate redistribution.

The remaining entries included sworn statements from members of the American syndicate and the Washington group. In general they agreed that a competent operator should average 3,000 ems per hour, but Miss J. Julia Camp, who had demonstrated Mergenthaler's machines at Baltimore and composed from dictation, thought that a first-class operator should be able to maintain 10,000 ems an hour for a full day's work after a 6-month training.

This advertisement was the first to refer to The Linotype Company and announced that the Linotype Company prospectus would be ready for issue on Tuesday 30 July at 10 am. The advertisement which ended with an offer of: 'Orders to view Machines in full operation at the Company's Show Rooms, at Blomfield House', was repeated on Monday 29 July.

The prospectus dated 27 July 1889 was published on Tuesday 30 July, and stated that the subscription list would open on Wednesday 31 July, and close at or before 4 pm on Friday 2 August, for town and the next morning for the country. The second statement modified the claim of savings by deducting the royalty of £80 per annum per machine to give a net saving of £13,020 per annum at the *New York Tribune*.

The main heading was a capitalised 6-column banner: 'The "Linotype" Company, Limited', over the statement that the company was incorporated under the Companies Acts, 1862 to 1886. The capital of the company was £1,000,000 divided into 196,000 shares of £5 each, and 1,000 Founders' shares of £20 each. Applicants for the £5 shares had to pay 10 shillings per share on application and pay the balance in three stages after allotment. A rider in small print stated: 'After a Dividend at the rate of 10 per cent per annum has been paid on the amounts paid up on all Shares, the Founders' Shares will be entitled to one-half the surplus profits.' Jacob Bright, MP was one of seven directors under the Rt Hon Lord Thurlow (chairman). No deputy chairman had been nominated and Joseph Lawrence was not named. The Prospectus stated: 'This Company is formed for the purpose of acquiring and working the patent rights for the United Kingdom originally owned by the National Typographic and Mergenthaler

Printing Companies, of New York, and now owned by the Linotype Syndicate, of London, for the composing machine or invention known as the Linotype.'

The usual list of advantages claimed for the Linotype was followed by further notes about the machines at the *New York Tribune*, a statement that shares in the American companies maintained a very high premium and remarks about the demonstrations given in London. This section reported that the Founders' Shares (carrying the privilege of prior issue of machines) had been acquired through the vendors by newspaper proprietors and members of publishing firms and others. Therefore, the Syndicate must have offered the Founders' Shares to prospective users of the Linotype on a privately selected basis; there is no evidence that they were offered to the general public.

Points raised in fifteen long lines of fine print at the bottom of the page included: an extrapolation from a count of the newspapers and general printers in the United Kingdom that 10,000 machines would yield a gross income of £800,000 per annum; the capital of the company would place £180,000 at the disposal of the Directors as working capital, which they considered amply sufficient for the manufacture of machines and for the general purposes of the company; and the vendors had fixed the price of the manufacturing rights at £820,000, payable as £600,000 in cash, £200,000 in fully paid up ordinary shares, and the Founders' Shares.

It was also reported that a contract dated 15 July 1889 existed between Jacob Bright, on behalf of the Linotype Syndicate of the first part, John Farquar Gilmore and Ernest Orger Lambert [Cottam's colleagues] of the second part, and Samuel Fyfe Easton on behalf of the Company, of the third part. This agreement showed that the promoters lay functionally between the syndicate and the Company. The sequence of negotiations, shown schematically in figure 38 links Hutchins to Bright, Bright to the Promoters, and the Promoters to the Linotype Company.

The prospectus included the statement: 'During the negotiations for the purchase of the property and the formation of the Company, arrangements have been entered into with various parties in connection therewith, and as to the preliminary expenses and subscription of the capital, to none of which the Company is a party. Applicants for shares must therefore be deemed to have **waived all rights** [my emphasis] to particulars thereof, whether under section 38 of the Companies Act, 1867, or otherwise.'

The prospectus ended with instructions for obtaining prospectuses and forms of application, and guaranteed that if no allotment were made the deposits would be returned in full. The Prospectus was reprinted next day. The final advertisement, filling the top half of the back page of *The Financial Times* on Saturday 10 August 1889, and repeated on Monday 12 August 1889 withdrew the original prospectus and announced a new one dated 8 August 1889. This ended the press campaign. There was no need for more publicity – the campaign had identified potential investors.

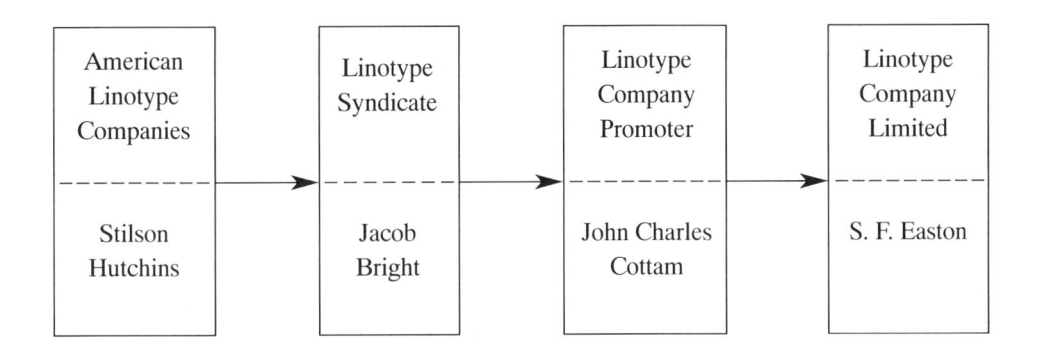

| American Linotype Companies | Linotype Syndicate | Linotype Company Promoter | Linotype Company Limited |
|---|---|---|---|
| Stilson Hutchins | Jacob Bright | John Charles Cottam | S. F. Easton |

Figure 38   Scheme of negotiations for the British rights in the Linotype

## 4   *Legal considerations in setting up the Linotype Company*

The prices for the manufacturing rights granted to Bright on 20 June were agreed between Hutchins, for the American companies, and Bright, for the Linotype Syndicate, on 11 July, just before the press campaign started. The deal was to be completed by 15 July at the offices of Mr P. G. C. Shaw, the purchaser's solicitor. The price for the British patents was £400,000, consisting of £266,665 in cash, payable in four stages, and 26,667 fully paid up £5 shares in the Linotype Company. The price for the patents for everywhere outside the United Kingdom and the Americas was £200,000 cash, payable in three equal instalments. The agreement stipulated that the Americans would not compete against the British in any of the areas covered by the agreement. They also offered technical support from The Mergenthaler Printing Company, other than Mr Mergenthaler himself.

This agreement signed by Hutchins and Bright was a requirement for the newspaper campaign to start on 13 July 1889. It would have given the Linotype

Syndicate a balance of £220,000 and Hutchins would have been entitled to $500,000 (about £100,000) in commission.

Details of the contract of 15 July (between Bright, Lambert and others) mentioned in the fine print of the prospectus were not revealed but obviously it set targets for the promoters to achieve to earn their payment. The Linotype Company, Limited was incorporated on 29 July 1889 as company number 29456; it had to be registered before the prospectus was published on 30 July. The memorandum of association named Bright and Lee Hutchins among seven subscribers, each taking one share; Lawrence and Stilson Hutchins were not included. The qualification to be a director of the company was a holding of 100 ordinary shares, with a par value of £500. Directors were to be paid £500 per annum with £1,000 for the Chairman and £750 for the Deputy Chairman.

When the subscription list closed, on 3 August, only £150,000 of share capital had been applied for. This was returned to the applicants on the advice of the Attorney-General, because the wording of the prospectus left it in doubt whether the Company could proceed to allotment with less than £200,000 working capital.

No copy of the new prospectus dated 8 August 1889, seems to have survived, but Mr Lionel Ashley gave some details about it at the statutory meeting of the Company held on 29 November 1889, four months after the company was registered. The new prospectus offered £100,000 of stock; £50,000 had been applied for, and the directors had decided to go to allotment on that amount.

A new Memorandum of Agreement dated 8 August 1889 was drawn up because Gilmore and Lambert had not met their obligations under the agreement of 15 July 1889. The Linotype Company would still pay the syndicate £820,000, but all in fully paid up shares consisting of £20,000 in Founders' Shares and £800,000 in £5 Ordinary shares; of which 40,000 shares representing £200,000 would not be entitled to any dividend for five years and were to be held by the Company and not issued by the vendors without the consent of the Board of Directors. The voting rights were also restricted to one half of the shares held by the vendor or his nominees. The memorandum was signed by Jacob Bright, for the Linotype Syndicate, before P. G. C. Shaw, Solicitor, and the Company seal was affixed in the presence of Directors K. Gillam and J. Bright.

The original agreement between Hutchins and Bright was cancelled on 15 August 1889 by adding two paragraphs to the document signed by them and witnessed by P. G. C. Shaw. Bright, the purchaser, undertook: 'to hand over to the said Stilson Hutchins all documents of title relating to the within mentioned premises other than the British patents.'

The above changes to agreements between the Linotype Syndicate and the Linotype Company made it necessary to alter the agreement between the American companies and the Syndicate, achieved by a Memorandum of Agreement made between Hutchins and Bright on 15 August 1889. The Linotype Syndicate agreed to buy the British rights for £500,000, in fully paid up £5 shares of which 25,000 shares would not participate in the distribution of dividends for five years; and the vendors and purchasers would not vote on half of their shares at any meeting of the company.

The revised agreement would have given the Linotype Syndicate and the promoters shares with a face value of £320,000. However, Hutchins had failed to restore American finances in the short term and could not expect any immediate payment of commission.

The final document dealing with the launch of the Linotype Company Limited was an indenture dated 28 November 1889 between Jacob Bright for the Linotype Syndicate and the Company, carrying a Stamp for £2,000 (paid for by the Company), allotting 1,000 fully paid up Founders' Shares and 160,000 fully paid up £5 Ordinary shares to the Syndicate in payment for the British manufacturing rights. This document was signed, sealed and delivered by Bright in the presence of P. G. C. Shaw and the seal of the Company was affixed in the presence of F. W. Gillam and Lionel Ashley, Directors, and Walter Thomason, Secretary.

## 5    *Public response to the launch of the Linotype Company*

The initial launch of the Linotype Company failed because both printers and investors had misgivings. Experts were intrigued by the mechanism but doubted its durability, criticised the printing surface, and objected to paying over twice the development costs for the manufacturing rights. The attacks on Cottam, the promoter, in the *Sun* and the *Hawk*, could have deterred potential investors. Calculations quoted in the prospectus, based on renting 10,000 machines assumed that, in time, all printers would use the Linotype, an unrealistic projection because no one achieves 100% market share. British printers found American type faces unacceptable and the American concept of rental without the option to purchase was alien to British marketing practices.

The attacks on the probity of the enterprise continued on 21 July, when the *Sun* recalled its article of 7 July and reiterated the list of failed companies. Its report was not completely negative; although criticising the inability to do varied work it described the machine as a masterpiece of ingenuity. The *Sun* condemned the purchase price of £820,000 for the patents for the United Kingdom as utterly absurd. The report inferred that the company could never earn satisfactory dividends with this millstone and noted that L. J. Jennings, MP, former editor of the *New York Times* had resigned from the board. Finally, on 17 August, the *Sun* claimed that the public had subscribed £80,000 in response to the first launch, estimated the cost of advertising at £20,000 and noted that the new prospectus offered shares to the value of £100,000.

Following its short comment about Linotype patents on 9 July, the *Hawk* began to attack Cottam in earnest on 30 July. It opened with the statement that: 'Mr John Charles Cottam and his gang seem to be fond of bubble promotions' and named several companies – shown in the lampoon of Cottam, see figure 39. The article pointed out that the same people had participated in most of Cottam's promotions and criticised the output from the Linotype, particularly letter spacing. The final comment was: 'I think I have said enough to keep any sane person from risking a sixpence in The Linotype Company, Limited.'

On 13 August, Mr Augustus Martin Moore editor of the *Hawk* wrote: 'I claim that there are very few persons who have saved the British public a million of money. Further, I contend that I have accomplished that most desirable end. My title to this distinction is founded on the fact that out of all the papers in England, and all the men who write upon them, the *Hawk* is the only paper and I am the one person who raised a vigorous protest against the launching of the Linotype Company.'

He criticised the second launch of the company ending a well-argued attack on the Linotype Company with the statement: 'And I now wish to inform the public that this article is written in defiance of the fact that Mr John Charles Cottam has threatened to prosecute me criminally for my previous article on the Linotype machine. I can only say that I heartily hope that he will do so, and will give me the opportunity of proving in open court which of us is the least interested in this discussion.'

On 27 August, under the title 'Promoters' Plunder', the *Hawk* referred to the contract of 15 July between Bright, Gilmore, Lambert and Easton. The article stated that Cottam and Lambert were partners, of 39 Lombard Street, Company

# AN OVERVIEW OF THE BRITISH LINOTYPE COMPANY

"THE HAWK." September 17th, 1889.

THE LITTLE GAME OF "CODDAM."

Figure 39  A lampoon of John Charles Cottam
*Source: The Hawk, 17 September 1889*

Promoters, and claimed that Cottam (a former telegraph clerk with a large railway company) and Lambert were to be paid: 'the startling sum of £200,000 and a portion of the founders' shares', for promoting the Linotype Company. The writer assumed that the partners would share about £180,000 after allowing for overheads; obviously he was unaware of the agreements between Bright and Hutchins.

The prospectus stipulated that applicants waived the right to know how the company was set up. The secrecy surrounding company launches at that time was scandalous; as shown by the following quotation from the end of the article: 'The public have no powers of finding out what becomes of their money, and I can only repeat that a change in the law should be made, compelling promoters to disclose the full nature of all contracts between vendor and promoter.' On 3 September, the *Hawk* reported that at the statutory meeting of the North British Water Gas Company (another Cottam company) the manager, Mr S. Fox, refused to divulge the amounts paid to 'certain persons' and told a shareholder: 'You can't have any knowledge of the work done by the promoter, and you can't have any right to know what price the persons who engaged him agreed to pay for his services, whatever these might be.'

There is no record of the total amount that the Promoters received from the Syndicate, but it was much less than the writer had estimated. A year after the company was launched Cottam and Lambert were allotted 850 ordinary shares reclaimed by the company from two female shareholders.

Moore had put Cottam in a position where he had to seek redress in the courts. The Linotype libel case opened at Bow Street on 22 August 1889. Moore appeared before Mr Vaughan charged as publisher of the *Hawk* with unlawfully and maliciously publishing a defamatory libel of and concerning Mr John Charles Cottam. Mr Besley, for Cottam, said that his client was objecting only to the passages that started: 'Mr John Charles Cottam and his gang seem to be fond of bubble promotions.' [Bow Street court reports prior to 1891 have, unfortunately, been destroyed.]

Mr James Maddock, a friend of Cottam, said that he had told Moore that he had no right to canvass for advertisements and then run the invention down; he continued that the defendant said that if Mr Cottam thought that £30 would shut his mouth, he was very much mistaken.

Mr Jacob Bright, MP, called by the defence on subpœna, gave some details of the two launches of the company, but declined to say what the syndicate gave for the patents. Questioned by Mr Besley he said that Mr Cottam could not make any profit except by the sale of the machines and their success. He considered Mr Cottam an honourable man.

Mr W. R. Horncastle, an advertising contractor, had joined syndicates to promote companies on the understanding that the advertising went through his agency. The £11,000 spent on advertising the Linotype Company was rather larger than usual. Cross examined by Mr Lewis he said the £11,000 included all the advertising and that Mr Lambert paid the amount.

Mr Besley said his only other witness was Mr Cottam and asked for an adjournment. When the proceedings resumed on 5 September, Mr Besley dropped the case without Cottam giving evidence. The defence lawyer, Mr Lewis, declared: 'I withdraw none of the statements that have appeared in the *Hawk*.' Bright was obviously uneasy under cross examination so it seems probable that he and Lawrence forced Cottam to drop the case.

Encouraged by this success, Moore continued to harass Cottam. In the next issue of the *Hawk* he wrote: 'Of course Mr Charles Cottam did not have the courage to enter the witness box at Bow Street on Thursday and submit himself to the cross examination he so much dreaded.' He hoped that Cottam would retire and: 'relieve the public from any more schemes such as Linotype, Moldacott, etc.'

On 17 September, the *Hawk* printed a copy of a letter (without names) sent on behalf of Bright implying that journalists could claim payment 'for services rendered to the Linotype Syndicate, by having written a notice ... relative to the Linotype machine.' On 3 December, the *Hawk* reported that Mr Justice Shirling was merciless in his denunciation of the infamous proceedings between Cottam and the directors of the Water Gas Company; the article continued with another attack on the Linotype Company.

At the end of 1889, the *Hawk* reported: 'Should Mr Cottam reappear, as, I presume, he intends to try, he will find I have several rods in pickle for him, and during the next year, I trust to be able to expose the doings of several other professional promoters and their colleagues about whom I have just as little to fear as I had of Mr Cottam.'

## 6  *Early British reactions to the Blower Linotype*

The Linotype Syndicate chose the most complimentary remarks from press reports of the machine in their publicity campaign. First reactions to the Linotype were mixed, and continued so for the next two years while the Blower was the only machine known to the British. Although some people considered the machine a significant breakthrough and were prepared to accept early shortcomings, others were determined to find fault. Probably the most significant complaints came from highly respected experts who compared the Linotype with the best hand setting and found it wanting.

Most critics thought that the machines were well built and an ingenious design, but criticised the quality of the output. They claimed that the time taken to decipher very bad copy would slow the speed of operation to that of hand composition; a valid point, but one that ignored the relative speeds when setting good copy.

Southward made typical comments in the *Printers' Register*: 'We perceived at once the remarkable ingenuity of the apparatus which, indeed, was self evident. But no opinion as to its economic features can possibly be formed while it is worked in a showroom by an expert ... We cannot, indeed, learn of anyone having seen the machine set ordinary copy for even one hour at a run.'

On 3 October 1890, he addressed the Balloon Society of Great Britain on 'Type-Composing Machines of the Past, Present and the Future', and claimed that the Linotype was a machine of the past. He said that the Linotype Company secretary had sent him a letter stating that it had ceased manufacturing and that an improved machine was about to be introduced, but he had been given no details. He also pointed out why the principle of the Linotype was essentially mistaken and impractical and went on to praise the Thorne typesetting machine, see figure 40. [Note that audio-typing techniques were used in 1890.] In the discussion several speakers praised the Linotype at the expense of the Thorne.

The discussion continued in letters to the *British and Colonial Printer and Stationer*. Southward repeated his case for describing the Linotype as a machine of the past. The secretary of the Linotype Company pointed out misleading statements in Southward's lecture, stated that new machines would soon be in England (implying that Square Base machines were on order) and claimed that Southward was paid by the Thorne Syndicate. On 16 October, the Linotype Company's solicitors accused Southward of making one-sided, misleading and untrue statements that in their opinion constituted 'actionable slander'. They called for a full retraction, otherwise they threatened to take legal action.

On Friday 24 October, John Place read a paper at the Balloon Society meeting, on 'The Linotype Machine, with its latest improvements'. He had seen the new machine in America and believed it to be a marvel of ingenuity. He said that Southward was justified in criticising the printing surface produced by

THE "THORNE" COMPOSING MACHINE WITH EDISON PHONOGRAPH ATTACHED.

Figure 40   The Thorne typesetter
*Source: The British Printer, May-June 1890*

hand-punched matrices from America, but said that new machine-made matrices accurate to 1-8000th of an inch, produced slugs that compared favourably with movable type. Southward told the meeting about the Linotype Company's letter and said that he would not make a retraction. He challenged the company to take him to court; he was prepared to stand or fall, by what he said. In an acrimonious discussion, he noted the comment in Reid's 1888 annual report, that 'no one of these machines has earned a dollar'. Mr L. Bright [Jacob Bright's son, Managing Director of the Linotype Company] denied that 'Whitelaw Reid had made a statement derogatory to the Linotype machine.'

Southward changed his opinion when he saw the improved Linotype and from then on gave the machine his full support.

## 7  *Progress of The Linotype Company*

The first launch of the Linotype Company failed for lack of investment. The revised prospectus offered shares worth £100,000 of which less than £50,000 was taken up. The directors decided to go to allotment on that amount. The British company had originally intended to pay the vendors £820,000 using the balance of £180,000 as working capital to manufacture and sell Blower Linotypes. With less available cash the directors invited tenders from manufacturing firms to build Linotypes under licence, but by November 1889, they had received no satisfactory quotations. The firms wanted orders for too many machines, at too high a price and would not agree to deliver machines in less than twelve months from receipt of order.

At the statutory meeting held on 29 November, the Hon Lionel Ashley said that the American company had told the directors that they had set up and equipped the factory in Brooklyn for £30,000 and thought that cheaper labour and materials in the UK would leave several thousand pounds for working capital. Therefore the Board had decided that the company should start their own factory; an incredible decision in view of the fact that they had originally budgeted for £180,000 to run the operation and went to allotment with under £50,000. It was reported that investors had applied for shares worth £50,000 but the Summary of Capital and Shares, made up to 13 December 1889, listed sales of only 8,944 ordinary shares with a face value of £44,720. The prediction of several thousand pounds of working capital showed that the Board expected no cash flow problems. They were gullible to take the figure of £30,000 at face value in view of the sworn statements by American users, cited in Linotype

Company advertisements, which showed that, despite vast sums spent on development, relatively few machines had been produced in the three years since the first Linotype was installed at the *New York Tribune*. The likelihood of long lead times and heavy overheads with no immediate return, could have been deduced from the tenders submitted by contractors. Although the Americans might have set up the Brooklyn factory for £30,000, because they had moved everything from Baltimore, including Mergenthaler's tools and many partly assembled machines, that figure would not have covered running costs until it was ready to deliver machines.

Ashley also reported that orders in hand included 90 machines for the *Leeds Mercury*, the *Scottish Leader* and the Carlyle Press Company managed by Mr William Burgess. [Southward was referring to Burgess when he claimed, in October 1890, that the Linotype was already obsolete.] Ashley hoped that the 90 machines would all be at work by Christmas or the New Year and that they would start to deliver British built Linotypes by February 1890. Most accounts state that 60 Blower Linotypes were actually imported.

Further confusion over the expected number of machines was provided by the *British and Colonial Printer and Stationer*, 23 January 1890, which reported: 'We understand that the Linotype Company have at present 120 machines on order, 40 of which are expected to arrive in England in the course of a few days, the rest to follow on completion.' In fact it was most unlikely that the Americans could have supplied 120 machines.

Ashley said that the Americans would supply machines, at cost price, to meet their immediate needs. Mergenthaler wrote that these machines had been sold for $1,000 each, but they had cost him $1,300 each to make and his price was below that of the Brooklyn factory. Therefore this price was a lure to dump obsolescent equipment. Had the Americans charged the full cost, the contractors' price might have seemed reasonable.

[In the 1894 report to shareholders Bright, as chairman of the Linotype Company, mentioned the supply of Linotypes from America but massaged the story to imply that they had been specimen machines placed on trial in three or four English newspaper offices: 'with a view to their paving the way to future orders whilst a factory was being established in England.']

In December 1889, the company leased a disused cotton mill in Manchester for £215 per year to set up a factory to make Linotypes. The optimistic forecasts of early production slipped badly and the factory had only just started to produce 'the rough and heavy parts of the Linotype' by March 1890. This was partly due to the nine-month lead time for Benton-Waldo punch cutting machines from Milwaukee, which cost £1,000 each, cash with order.

By June 1890 the company was in dire straits; the cost of setting up the factory was much higher than predicted and, as it was difficult to raise more capital, Joseph Lawrence, now Deputy Chairman, met the American company in New York and negotiated a reduction of £450,000 in the ordinary share capital. There does not seem to be any record of how this was divided between the British syndicate and the American companies. In July 1890, during this visit, Lawrence saw the improved Linotype and cabled England to stop Blower production immediately. It is evident that developments were kept strictly secret within the American companies. Peter Whittaker (see biographical notes) claimed that his grandfather was furious that the British had been misled into thinking that the Blower was the definitive Linotype because Matthew Whittaker was quite unaware of the new machine during his training in America.

On 27 August 1890, the company held an extraordinary meeting of shareholders to authorise the cancellation of £450,000 ordinary shares, negotiated in New York by Joseph Lawrence, and authorise the creation of £100,000 of 6% Preference shares. Jacob Bright, who presided, said that when the new shares were issued the actual capital of the company would not be increased because £137,000 of ordinary capital was unissued. He hoped to announce the reduction of ordinary share capital to the annual general meeting at the end of the year. The preference shares were offered 'with a bonus upon the first £50,000 of 50% in Ordinary Fully Paid Shares.' The chairman reported that 60 machines had been imported from the USA, but British newspapers, with the exception of the *Scottish Leader*, would not use them until they were supplied with English-faced type. He claimed that it would be possible to make Linotypes for £100 each and said that an improved Linotype had been produced in America. It is not known how he arrived at the figure of £100 per machine; it sounds like the marginal cost of labour and materials. A realistic price would have included overheads, particularly the cost of manufacturing rights, which would be lost when the patents ran out. It was expensive to stop Blower production before any machines were made, but not as costly as it would have been to cancel outside contracts. It meant scrapping all construction to date, adapting existing equipment and ordering new plant and tools.

There was friction between the British and the Americans. An agreement, dated 21 December 1890, referred to a payment for outstanding accounts up to 31 December 1890. The Americans claimed $45,000 (£9,000) and the British counterclaimed for repairs. The Americans agreed to accept £7,000 ($35,000), in the form of 1,400 fully paid up 6% Preference shares.

The first 'improved' Linotype imported to the UK was erected at The Economic Printing and Publishing Company, Worship Street, Finsbury, for Machen to demonstrate at the adjourned second ordinary general meeting on 27 February 1891. The demonstration was a complete success. The machine was moved to the Printers' Exhibition at the Agricultural Hall, Islington on 16 March 1891.

## 8  *The Blower Linotype in the United Kingdom*

Most historical accounts state, erroneously, that the first Linotype installed in Britain was at the *Leeds Mercury* in 1890 although it may have been the first English newspaper to use the machine. The first British installation that

Figure 41   Battery of Blower Linotypes at the *Sheffield Telegraph*
Source: Leng, How we publish our papers, 1892, p 15

gave demonstrations of Linotypes and used them to set Lawrence's railway magazines was in Stilson Hutchins's offices at 37 Southampton Buildings. These machines were supposed to have been moved to the Linotype Company's new offices in New Broad Street, but they might have gone to Mr Burgess's Carlyle Press, which had the contract to print the railway magazines. In 1889 these journals carried notices that they had been set by Linotype.

Early orders came from provincial papers, where there were fewer restrictions than in London, but it took time to fill them. As noted in the previous section, most British printers did not like American type faces and were prepared to wait until English matrices were available.

Figure 41 shows a battery of Blowers and a Square Base Linotype at the *Sheffield Telegraph* in 1892. This seems to be the only contemporary illustration of a battery of Blowers to be published. Six machines can be seen, but it was reported that the newspaper had eight Blower Linotypes. No contemporary photographs of Blower Linotypes have been found in either Europe or the USA.

# EIGHT

## The Linotype – a technical summary

This chapter deals with technical aspects of early Linotypes that were not relevant to the main biography. It aims to show how each new machine overcame drawbacks of its predecessor until the Simplex model provided a design that endured, with enhancements, until hot metal was superseded by computers. It also discusses early problems of making matrices, the requirements for casting good slugs and routine maintenance.

The biography has shown that Mergenthaler became obsessed with developing the Linotype. As a watch maker he was trained to handle machinery that did not wear out quickly, unlike the Linotype which was subject to external stresses and easily damaged. His designs may have been influenced by current trends in engineering and printing, but he deserved most of the credit for designing the machine that started the hot metal era.

Mergenthaler experimented with several technologies before producing a practicable machine. From 1878, when Clephane first suggested making a stereotypic mould by impressing characters into papier-mâché, every model was designed to produce lines of type. Before the introduction of direct casting it took at least two processes to produce a printing surface. The main advantage of the hot metal machine was that the line of type was produced by a single operator. This made it superior to all contemporary machines that set movable type.

### 1  Single operator linecasting machines

Lines of type were produced in two phases, manual typesetting followed by automatic operations to: justify and cast the line; eject the slug; and distribute the matrices, during which the operator could start to set the next line, as shown diagrammatically in figure 42.

Mergenthaler first demonstrated the second band machine at his Bank Lane shop. It cast up to four lines per minute, marked the start of the hot metal era, the end of basic research and the start of the development phase.

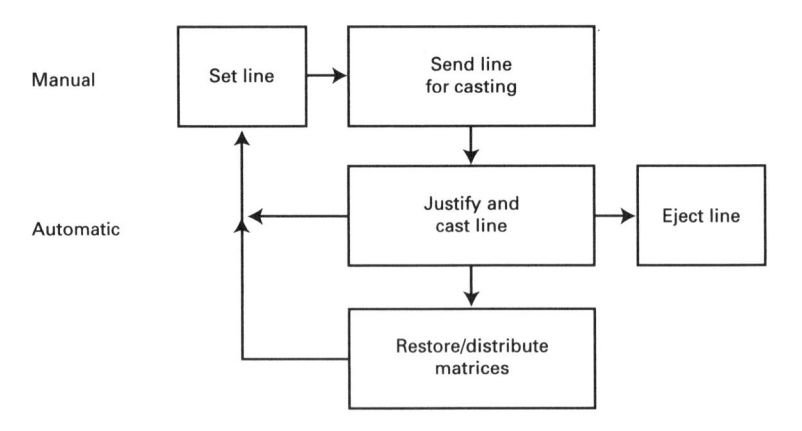

Figure 42   Cycle of operations for producing lines of type

It also demonstrated the following features, common to all production models of Linotype:

1  Direct casting of lines of new type,
2  Operation by one person,
3  Elimination of the chore of distribution at the end of the job,
4  Continuous operation.

The bands were tapering wedges; long isosceles triangles carrying a full set of characters, including spaces, in order of width. The bands were hung alternately from their broad and narrow ends to leave no gaps for metal to penetrate. Although lines were cast at the same place as they were composed the operator could set a new line during casting. Each keystroke set a pin to stop the band where required. A spacing device remembered the location of spaces. When the line was complete the operator calculated the spacing needed to justify it and entered that number to set the spacing stop pins. The bands were lowered to the stop pins, the line was locked, the pins were retracted and the line was cast. The operator could resume setting as soon as the pins retracted.

Despite these advantages there were several major drawbacks:

1  It was not possible to detect or correct errors made while setting a line,
2  It was not possible to tabulate accurately,

3   It was not possible to set characters that were not on the band and, as each band carried a full alphabet, some characters would be little used,
4   The machine had 58 bands, all of which had to be used on every line, which meant that all lines had to be the same length.

The second band machine, with automatic justification, was exhibited at the Chamberlain Hotel in Washington, DC, in February 1885. It was originally intended to be the production model, but Mergenthaler assured his backers that a machine that used circulating individual matrices would overcome the above problems and should be developed.

## 2   *The individual matrix Blower Linotype*

The production model of the individual matrix machine, later known as the Blower Linotype, see figure 43, performed all the functions of later models, but with a radically different configuration. It was essentially a prototype but more effective than the second band machine because:

1   The machine could carry an appropriate number of individual matrices of each character, as in a type case, and the matrices of each character would be subject to similar usage,
2   Lines could be checked and corrected in the assembler, and limited tabulation was possible,
3   Special characters could be inserted by hand.

The Blower Linotype was more expensive than originally planned. The $1,300 cost of each machine was more than three times the price of $400 agreed by Hine and Mergenthaler as the ceiling for commercial success.

To understand the concepts of Linotype composition and to show how later models were superior to the Blower, it is necessary to describe the sequence of operations in some detail:

1   The matrices were held in a magazine made up of vertically mounted brass tubes of rectangular cross-section, designed for one size of type and not suitable for any other. The tubes, which became shorter from left to right, held between eight and fourteen matrices. In the patent specification Mergenthaler claimed this to be an advantage because the shorter tubes would hold fewer matrices of less used characters; so there would be a smaller drop and less chance that matrices would be damaged during distribution. This implied that the keyboard

had been designed to put the most used matrices to the left of the magazine,

2  There were 107 typewriter-like keys on the four row keyboard. The escapement mechanism, shown in figure 2 of figure 43, was meant to deliver one matrix at a time, but the keyboard was completely manual so results depended on the skill of the operator. Several matrices might drop if the keys were struck too hard, but no matrices would fall if the touch were too soft. The original keyboard layout is unknown, but diagrams from early patents suggest that it was probably alphabetical. Mergenthaler refused to adopt the typewriter 'QWERTY' arrangement because it would have been ineffective. When users complained that the keyboard layout was inconvenient he devised the 'etaoin/shrdlu' arrangement which put the most used letters in the leftmost tubes and therefore distributed them first. In 1889, the *Scientific American* stated that there were alternate keys for the most used letters and several compound characters (ligatures) on the keyboard. Each matrix carried a single mould; there were no italics or small caps on the keyboard, but the operator could insert special characters as extraneous sorts. When a matrix dropped from its tube it was carried down a pair of inclined wires by a blast of air and helped into the assembler by a jogging action that held the matrices in place,

3  Words were separated by 'space bands'; see figure 3 of figure 43, double wedges that became wider when the longer wedge was raised, held in a 'space box' to the left of the keyboard. Assembled lines had to be slightly loose to justify effectively. The assembled line was moved left into the elevator after which the operator could start to set a new line,

4  Further operations were controlled automatically by a set of cams that turned round a vertical axis. The top of this battery of cams appears as a conical shape at the left rear of figure 43. After journalists described this mechanism as the brain of the Linotype, some people actually believed that the machine could think. The line was lowered to the level of the metal pot and held between two vice jaws. Then a 'justification bar', actuated by a heavy weight, raised the long wedges of the space bands to spread the line until it was tight. This ensured that all the words in the line were equally spaced. Then, with the metal pot, mould and justified line pressed tightly together, molten metal was pumped through the mould and on to the matrices to cast a line of type. The machine could set up to six lines per minute, so the total time from casting to ejection was under ten seconds, by which time the slug had to be solid. Compressed air was sometimes used to cool the mould,

5  The mould was attached to a 'mould wheel' which turned through 180° after the line had been cast. During rotation, knives behind the wheel trimmed the

Figure 43  The Blower Linotype with detail of magazine tube escapement, space band, matrix and slug
*Source: Engineering, 28 June 1889, p 719*

slug to the required height for printing. An ejector blade then pushed it past parallel knives, which trimmed it to the required width, into a vertical galley at the front of the machine. The slug could be leaded automatically (to increase the space between lines of text) by using a mould with a larger body than the type face in use,

6  As the mould wheel started to turn, the elevator raised the matrices and space bands to the top of the machine where they were separated for distribution. The space bands, which had plain tops, dropped down a long chute into the space box. The matrices were fed on to a toothed distributor bar, similar to that shown in figure 44, and pushed forward by a system of moving combs mounted on an endless belt. Seven symmetrical ridges were cut into the distributor bar, which allowed for up to 127 binary combinations of retaining teeth. The combinations of teeth on the bar matched those on the matrices, which ensured that each matrix was distributed automatically into its proper channel.

Mergenthaler devised an electro-magnetic brake to stop distribution if the matrices did not distribute properly, but it used batteries and was not very reliable. The air blast was also unpopular. Most firms installed a large steam or gas engine in the basement to provide power by means of belts and pulleys. Some early illustrations of Linotypes show the flues used to carry fumes from gas heated metal pots. Mergenthaler's experience with the first order of twelve

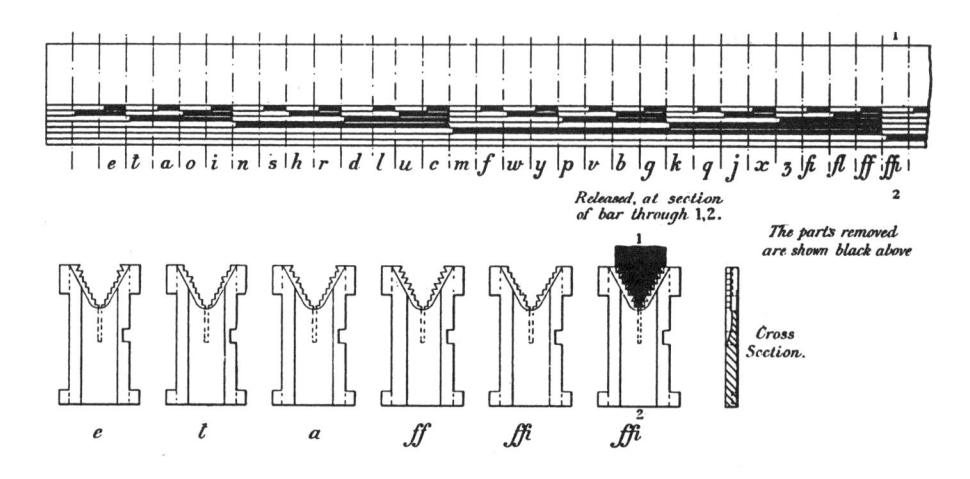

Figure 44   The distributor bar and matrices of a standard Linotype
*Source: Legros and Grant, Typographical Printing-surfaces, p 425*

machines highlighted many problems in the original design, but he was not allowed to modify the basic pattern.

In his 1889 report Reid stated 'that it had recently been deemed best to suspend all work on parts beyond the 200.' He accepted the principle of the machines but felt that refinements, which he had previously refused to sanction, were possible. The defects listed in his 1888 report were struck out of the draft of his 1889 report, probably because Abner Greenleaf had given him a glowing account of Mergenthaler's new machine.

## 3  The Square Base Linotype

This model introduced the typical Linotype shape used on all subsequent machines. It looked so unlike the Blower, see figure 11, that it is relevant to describe its structure and casting cycle, to show that both machines carried out the same operations, in the same order:

1  The separate vertical tubes that were unique to each size of type were replaced by a one-piece inclined magazine with 90 character channels of about equal length, suitable for a range of type sizes. The matrices were modified so that the lugs of each character matched the thickness of its magazine channel, regardless of type size. Each channel could carry 20 matrices, sufficient to set most lines, unlike the Blower which had extra tubes of the most used characters but even then sometimes ran out of matrices (known as short fonting). The sloping magazine increased the height of the machine to about seven feet; so the operator needed a step to reach the distributor,

2  The keyboard was completely changed. The 'etaoin/shrdlu' sequence was retained but the keyboard was arranged in six rows of fifteen keys grouped in three equal blocks of 30 characters: lower case on the left; numerals, special characters and punctuation marks in the middle; and capitals on the right. The keyboard was power driven. When the operator touched a key lightly a cam dropped on to a rotating rubber roller and raised a rod that released a matrix from the magazine. Like the Blower, this machine used single character matrices,

3  The air blast and wires that guided the matrices into the assembler were replaced by a moving endless leather belt inclined at about 45° to the vertical. The matrices fell from the magazine through a guide plate on to the leather belt and were fed into the assembler by a rotating curvilinear triangular wheel that replaced the jogging mechanism. The space box was redesigned so that the

space bands dropped into the assembler from the right-hand end of the box instead of from the left,

4  Lines were sent for casting by raising the assembler box until it tripped a mechanism that moved the line of matrices and space bands left into the first elevator which then descended to the casting position. The operator lowered the assembler box to resume setting. The casting process was similar to that of the Blower Linotype, but the mould wheel turned through three quarters of a circle (270°) before the slug was ejected. Therefore, slugs entered the galley vertically instead of horizontally, the extra 90° turn gave a little more time for the slug to harden before ejection, and the mould wheel turned through only 90° from ejecting one line to casting the next,

5  Two elevators replaced the single elevator of the Blower. After the line had been cast, the first elevator rose to its highest point and a second elevator carrying a short fully toothed rail, that matched the distributor bar, swung down from the top of the machine to rest beside the first elevator. A lever pushed the matrices and space bands towards the second elevator where the matrices were threaded on to the toothed rail to be raised to the distributor. The space bands, which had no teeth, were left behind and returned directly to the space box. This simplified the mechanism of the space box and reduced the chance of a line running out of space bands. It also avoided the problem of space bands sticking in the long chute from the distributor. The matrices were raised to the top of the machine and fed on to the distributor bar one at a time. The shape of teeth on both matrices and distributor bar were changed and the fragile combs of the Blower were replaced with a rugged system of revolving screws that carried the matrices along the distributor bar under the control of a mechanical brake,

6  The vertically mounted cams of the Blower were replaced with a horizontally mounted set. The justification mechanism was still actuated by heavy weights.

The Square Base was so much better than the previous model that the Mergenthaler Printing Company stopped supporting Blower Linotypes as soon as the new machines became available. Further, the British Linotype Company abandoned production of Blowers before a single machine was completed. Although some Blowers remained in use for a few more years, most firms changed to the improved model as soon as possible.

The most significant features of the Square Base Linotype were the second elevator mechanism; increased capacity of the magazine; flexibility in choice of font size and the light touch of the new keyboard. Blower operators ridiculed this as button tapping rather than operating. The two greatest drawbacks were the weight-actuated justification mechanism, that could be sluggish, and the massive square base. The new magazine was a mixed blessing; it provided increased capacity and an easier way to change fonts, but could only be removed from the back of the machine. This was a two-man job and dead space had to be left behind the machine to give the men room to work. Despite the radically different appearance of these three machines, both the Blower and the Square Base Linotypes performed the same cycle of operations, in the same order, as the Second Band machine.

## 4 *The Simplex Linotype*

This was the definitive model, shown in the *British Printer* advertisement of figure 28. Although the Linotype was regularly enhanced during the hot metal era, all subsequent models conformed to the same basic design. The first model of this Linotype was built in February 1890 and displayed at the Judge Building in New York, but did not come into regular use before 1893 in the USA and 1895 in the United Kingdom.

In time this machine became known as the Model 1, but the name can be misleading because the terms: Square Base, Simplex and Model 1 are sometimes treated as synonymous. The sequence of operations on the Simplex was the same as on the Square Base machine, but there were several improvements. The keyboard was given a lighter touch; the justification mechanism was made more positive by using powerful compression coil springs instead of weights; and the base was made lighter by introducing the star (or claw) pattern used on most later models, which made it easier to fit the justification springs noted above. No other new models went into quantity production during Mergenthaler's lifetime.

Early Linotypes were designed for body sizes from 5 point (pearl) to 11 point (small pica) and measures of up to 30 ems, although the early Blowers were limited to 24 ems. At that time much legal work was set in 12 point (pica) in lines up to 42 ems wide. It was not really satisfactory to butt two 21 em lines together because spacing across the two halves of the line might not be consistent. Therefore, printers 'pressed for a machine to set their type matter.' This resulted in the Pica machine, a Simplex that could handle fonts up to 14 point (english) and measures up to 42 ems.

## 5  *Significant enhancements to the Linotype*

Possibly the most obvious shortcoming of early Linotypes was a lack of flexibility. As a machine for narrow measure newspaper setting it was far ahead of its rivals but it took over an hour to change the Blower from news work to set *The Tribune Book of Open-Air Sports*. Both American and British inventors contributed many significant improvements that were standard features of later models. The two most important enhancements were the double-letter matrix and the two-line letter, which Reid had called for in his 1888 stockholders' report.

The double-letter matrix was invented by Dodge, who applied for a patent in 1892 which was granted in 1895, but it was not announced until 1898 after it had been thoroughly tested. Dodge had been President of the Mergenthaler Linotype Company for about seven years by then, so he must have decided to delay the announcement. By putting two dies on each matrix, as shown in figure 45, the operator got two alphabets from one set of matrices – roman with either italic or boldface. This feature, which could easily be fitted to existing machines, was advertised widely and welcomed as the most important advance since the Square Base superseded the Blower.

The operator was able to access either font by using the auxiliary shelf (or italic rail). This was a two-part slide fitted to the assembler box which was pushed forward for the alternative font and retracted to return to the basic type face. There were grooves in the line transport mechanism, the first elevator and the mould to keep the matrices in position during casting, even when lines were repeated. The matrices dropped to the lower position when the line was distributed.

Dodge had several other patents to his credit, most aimed at improving the quality of the printing surface. These included modifications to the casting mechanism and the metal pot.

The second important invention was the two-line letter for printing drop letters at the top of small advertisements. Mergenthaler invented one version in the USA and Matthew Whittaker produced another in the UK. A British Linotype Company report showed that Whittaker's version was invented first (British patent 10257 of 1894).

Inventors introduced new devices to make machine changes quick and simple. The original moulds were solid pieces of carefully machined metal made for one measure and one body of type. Printers who used several sizes of mould saved costs when the universal mould was introduced. This consisted of four pieces: a base screwed to the mould wheel, a removable cap and two mould liners. These liners were distance pieces, accurately machined to standard body sizes, that fitted into slots in the base of the mould to set the size of the slug. Right hand liners were of standard length and varied only in thickness, so only one was required for each type size. Left hand liners which varied in both thickness and length were used to set the measure of the slug. This brought down the expense and bulk of equipment needed to set a range of type sizes and measures.

Large bodied slugs (greater than 12 point) used a lot of type metal and cooled more slowly than less bulky ones. This led to the invention of the recessed mould which combined size with strength yet saved metal.

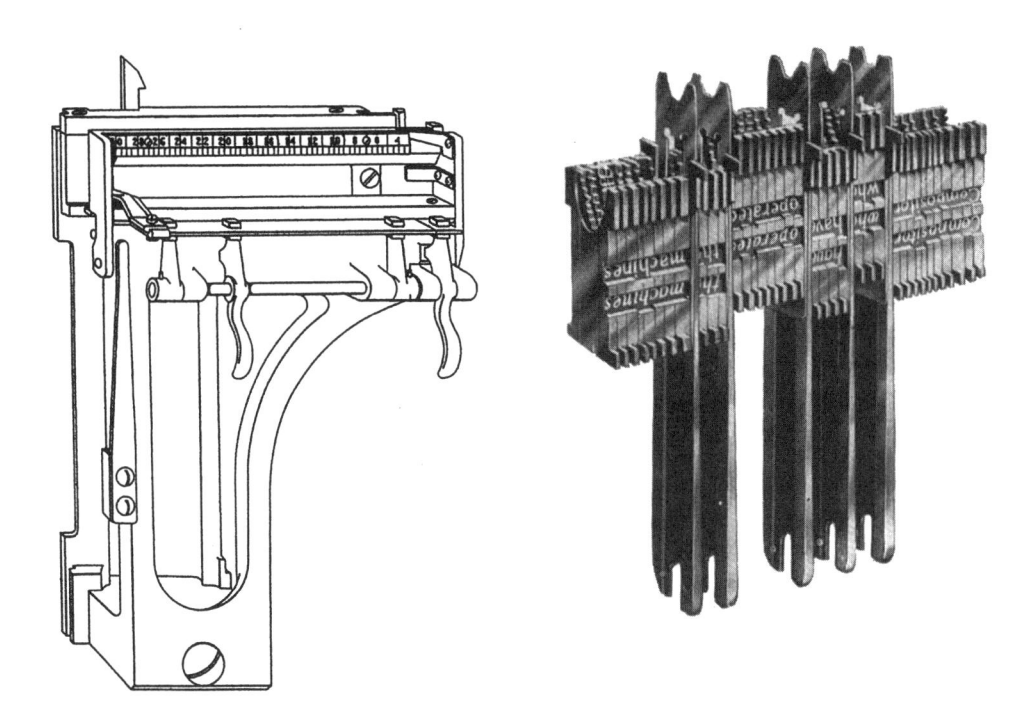

Figure 45  Assembler box and spaced line of double-letter matrices
*Source: Linotype Company documents*

The first mould wheels were designed to take just one mould, but were soon modified to take two or four, placed evenly round the wheel. The operator could swing the wheel round by means of a handle to change quickly from one mould position to another.

To make a complete change of measure and body size it was necessary to adjust or change several other elements including: the matrix magazine, the parallel trimming knives and the ejector blade.

The adjustment of the trimming knives was simplified by means of the universal knife block. The knives were reset by moving a lever to the required width, shown on a scale. Ejector blades were inserted by hand in early Linotypes; they were metal plates inserted through the mould and held by locating studs. They were all of equal length, but the width varied with the length of the slug being cast. In early machines the thickness of the blade was also chosen to match the body of the slug but, as casting technology improved, this became less critical. Setting the ejector became simpler with the introduction of the universal ejector which consisted of a set of narrow slices of blade, the lowest being four ems (about two thirds of an inch) wide and each of the others two ems wide. The operator simply selected the appropriate ejector size.

On 28 July 1891 Hine asked Mergenthaler, in confidence, to produce an automatic metal feeder before someone else patented the idea. It had been discussed at least two and maybe even four years before. This device would save the operator from having to leave the keyboard to feed the metal pot by hand and as it added extra metal gradually would also give more consistent metal temperatures and better slugs. The British Linotype Company obtained a patent on one form in 1898 and Dodge was granted a British patent on another version in 1901.

## 6  *Constraints on materials*

The Linotype was an impressive looking machine of which only a small part actually produced lines of type. In the main it was a delivery system that assembled matrices to make the mould of a line and distribute those matrices after the line had been cast. The machine was worthless if its slugs were bad. The difficulty of making good matrices held up delivery of the first Linotype to the *New York Tribune*. Reid wanted the print from his machines to match the type used on his paper. The first matrices were made by inserting nickel electrotypes of movable type into brass mounts. They failed because the nickel plating peeled off.

In early February 1886, Reid was infuriated when Mergenthaler, the perfectionist, was so disgusted with the results of electrotyping that he abandoned this method and started experiments to make matrices by a casting process. This should have been cheaper and better than electrotyping, but called for an alloy with special properties. It had to repel type metal and must not deform under heat or pressure. The problem was that during the line casting process, molten metal was being forced against the matrices while the space bands were forcing the line tight against the vice jaws.

In mid-May 1886, Mariner and Co of Chicago, Chemists and Assayers, submitted a sample of metal to be considered for casting matrices. The requirements presented to them were that it should be hard; should not be brittle; should be a fair conductor of heat; should present no affinity for type metal; should be highly fluid when molten; and should hold its body without shrinkage or expansion in the process of cooling. It was plain that any metal which would operate satisfactorily must have a higher melting point than type metal. They suggested about 850°F for a metal for making matrices by casting, rather than engraving. The proposed alloy containing 77.2% zinc, 17.0% tin and 5.8% copper was not used. The casting method was dropped in favour of hand cut steel punches, which were superseded by precise steel punches cut on Benton-Waldo equipment. These were used to impress dies in hardened brass.

When Reid complained that the Blower produced poor slugs the requirements of Linotype metal were not fully understood. It had to have a low melting point; run freely into the mould; solidify in under 10 seconds; expand on cooling to give a sharp printing surface and have the durability to make a stereotype or a direct print run. Such an alloy should contain lead for bulk; antimony for hardness and expansion on cooling (although it will raise the melting point); and tin for toughness and to help the lead and antimony to mix. Early alloys based on the metal used to make movable type contained too little lead to cast good Linotype slugs. In the early days it was not realised how much the metal contracted on cooling. Linotype slugs mixed with hand setting tended to be lower than movable type and would not print properly. Metal suppliers noted that Linotype metal was necessarily softer than other printing metals and recommended an alloy with 4 parts tin to 11 parts antimony and 85 parts lead. This mix was completely solid at 463°F, completely liquid at 477°F and reasonably durable. [It was difficult to control the temperature of gas-heated metal pots which were very sensitive to draughts.]

The Linotype mechanism put constraints on type face design because the machine could not emulate all the features of manuscript; in particular, where adjacent characters overlap vertically. In movable type the character was cast on a narrower body than itself; any projecting part that rested on the body of an adjoining type was called a 'kern'. This occurred most frequently in combination with the lower case letter f. The ligatures fi, fl, ff and ffi were used in British newspapers during the hot metal era. The type designer needed to balance the slope of the italics on a matrix against the overspaced appearance that it produced. In September 1898, the Linotype Company dismissed the problem by announcing the death of the kern in a short comment in *Linotype Notes*. The invention of the double-letter matrix magnified these problems; the designer had to find a workable compromise between italic and roman characters on the same matrix. In practice there would be trade offs between both fonts; the slope of the italic face would have to be balanced against an overspaced or expanded companion roman face. However, printers could order special ligatures as extraneous sorts which meant assembling matrices by hand, quite contrary to the philosophy of the Linotype.

Users soon complained that standard space bands were too thick for small faces like agate and gave an ugly over spaced appearance. As a consequence the company produced three sizes of space band: narrow, regular and wide. Newspapers generally favoured narrow spacing.

## 7  *Maintenance and operation of the Linotype*

The Linotype was a rugged machine that, with reasonable use, would last for decades. In fact, the Model 1 was so popular that there was an outcry early in 1953, when the British company announced that they would stop supporting the machine after 1 June 1953. At that time all surviving machines were over fifty years old. After 'a long and friendly discussion' on 23 June 1953 the company agreed to continue to supply spares until July 1958. The basic maintenance procedures suggested in this section were intended for the average operator with no special engineering skills.

Correct temperatures were vital; optimum metal pot temperature was about 550°F, which could be checked roughly by plunging a folded piece of paper into the molten metal. If it charred slightly the temperature was reasonable. If the metal pot were too cool it would cast imperfectly. If the temperature in the main part of the pot became too hot tin would burn off. A lack of tin would cause the metals to separate and possibly result in the throat of the metal pot becoming choked with antimony.

Even in a carefully run installation a certain amount of dirt and dross could contaminate the metal pot, therefore the operator had to carry out the following rudimentary maintenance, preferably at the start of each shift. Remove the pump plunger from its well and scour it briskly with a wire brush to remove any dross; loosen any dross from the sides of the well with a suitably shaped scraper; finally, scrape generally round the metal pot to remove traces of dross before they could build up. Skim off filthy dross, but not to the extent of lowering the tin content. Further, leave a little clean surface dross to protect the metal from oxidation.

No dirty metal should enter the metal pot, therefore used slugs were cast into new ingots. This was a task for an expert; if the dangers of burns and poisonous fumed were ignored the harm done could exceed the marginal cost of exchanging used slugs for new metal from the foundry. The operator had to avoid adding too much metal at once, thus chilling the metal too much to produce good slugs, making the level too high and possibly getting splashes, or letting the metal level drop so low that the slugs would be hollow.

If the mouthpiece of the metal pot became clogged, the operator could try to clear it gently with a wire probe. However, if the problem persisted and the mouthpiece had to be drilled out or needed specialised attention, it was advisable to call an engineer. Similarly, any adjustments to the seating of the pot which called for mechanical skills, was best left to the engineer.

Space bands had to be cleaned at least once per shift. The metal pot always cast against the same point of the shorter part of the space band and, since there was a maximum of 30 space bands in the machine, they were used repeatedly. At every cast it was possible that a trace of metal would stick to the space band. This deposit could be removed easily with a razor blade; if it were not removed it would grow and, in time, squash into the walls of neighbouring matrices. This would produce hairlines in the setting and, worse still, damaged matrices would corrupt sound ones until finally the whole font would have to be replaced. It is evident that this was not appreciated in Mergenthaler's day because he noted on 3 January 1894 that the machines in a Philadelphia newspaper were clean apart from the spaces being full of lumps. In time those lumps would have crushed the walls of any matrix, even ones made of a durable material like steel. A possible disadvantage of using steel for matrices was that sliding friction might cause wear on expensive parts, like the distributor bar; this would be less likely with matrices made from a softer material like brass.

Space bands had to slide easily for the justification mechanism to work properly. A sticking space band would stop the process of justification before the line was tight, which would cause a splash, probably break the space band and damage the matrices. Bands had to be lubricated every shift by polishing sparingly with graphite dust and wiping off any excess before the bands were returned to the space box. Graphite entering parts of the machine lubricated with oil could clog the mechanism with a muddy deposit.

For good slugs, both matrices and space bands had to be used properly. Space bands were inserted with the shorter wedge to the right, as viewed by the operator, because the bands were made fractionally wider on the side facing the mould to ensure a tight fit during casting. Further, no line should be sent away with two space bands together which could leave openings for metal to enter during casting, so causing a splash, damaging the matrices, bending the space bands, and stopping the justification mechanism from working properly.

A range of lubricants was required, from heavy oils to the finest grade of oil and graphite. Some parts of the machine were quite rugged, others delicate. To avoid contaminating parts of the mechanism with the wrong lubricant, the basic advice was: 'Keep the machine clean and free from rust and traces of type metal, lubricate sparingly and clean off surplus lubricant that might foul neighbouring parts of the equipment.' Some users thought that the magazine should be lubricated with graphite but this was not recommended. The magazine could be cleaned occasionally with a long magazine brush, to remove dust and dirt that entered through the distributor. Some early users suggested washing matrices in benzine but, unless they were thoroughly dried and polished, the benzine acted as a flux that made type metal stick to them.

Much routine maintenance was a matter of common sense, but in general the operator was not trained to make mechanical adjustments. When they were necessary it was usually cheaper and quicker to call for a qualified engineer to service the machine.

# NINE

## *Facts and fancies*

Before the Linotype Company was extensively advertised in the United Kingdom little was known about the Linotype in the USA. In fact, the *Inland Printer* published no article about the Blower Linotype until six months after similar pieces had appeared in the UK. It even illustrated its article with the same wood cut as the British had used in the summer of 1889. Hence most contemporary accounts of the Blower Linotype came from Britain rather than America. The weight of opinion of the first commercial model was that it was ingenious but fragile and needed radical modification before the trade would accept it. For some perverse reason the press, and people who should have known better, went out of their way to concoct tales about the inventor, his machine and how it got its name. This chapter aims to debunk fictionalised accounts which often obscure more remarkable facts!

### 1   *Contemporary comments about the Blower Linotype*

The first popular technical description of the Linotype which appeared in the *Scientific American* on 9 March 1889 may have been produced as a preliminary to the campaign to sell the manufacturing rights to the British, see figure 8. It was a modest account that made no extravagant claims. The article in the 19 May edition of the *New York Tribune* had a different purpose, see figure 12. It was pure propaganda intended to show how the early problems with the Linotype had been solved and how it was making profits for the paper.

On 21 June 1889, the *Railway Press,* from which figure 36 was copied, published one of the first articles about the Linotype to appear in the UK. This four-page item praised the machine, not surprising since it came from the office of Stilson and Lee Hutchins who were trying to sell the manufacturing rights to a British syndicate. In flowery language it stated that: 'After seeing a line of type set up . . . an observer will have no difficulty in believing implicitly in the absolute veracity of the wonders recorded in the Arabian Nights of Aladdin and his wonderful lamp. It is fortunate for Mr Mergenthaler and his co-inventor, Mr Clephane, that they did not live in the earlier days of printing, as they would have stood a very fair chance of being burnt at the stake as wizards or professors of the black art, and with much greater show of reason than was usually the case.' It also claimed that Mergenthaler had named the Linotype and that Stilson Hutchins had brought the first two machines to England.

As mentioned in chapter 7, John Southward was somewhat less upbeat. During the summer of 1889 he received letters about the first model of Linotype from two people who were disappointed with the machine. The first, dated 9 July, from William Blades (see biographical notes) noted that he was greatly interested and surprised with its wonderful mechanism but did not think it would put many compositors out of work. Its best work was only fit for poor newspaper work and utterly unfit for good book work. Its speed when running straight on was certainly rapid but was greatly reduced when setting anything other than roman; corrections were done more easily than in any former machine.

The second, dated 14 August, was from H. S. Bishop who had made a hurried visit to see the Linotype machine. He was charmed by its ingenuity but thought it required improvement before it could be used extensively in practice. He thought the distribution was the best feature, but thought that it could be improved and the matrices could be very much better. He saw they were not true and square when examining them before composition started and was not surprised to see some two letters close together while the next two appeared to have a space between.

The *Printers' Register* was scathing about an early circular issued by the Linotype Company and asserted that: 'The style of the printing is the fullest corroboration that we could have wished of the truth of our severest strictures. It is simply execrable – and the veriest "cock robin" shop would be ashamed of it.'

Eb Williams who had been connected with Linotype operation for three years at the *Chicago Daily News*, wrote about the machine for the *Inland Printer*. While accepting that it would probably play an important part in the future production of reading matter it could not and never would be able to produce first class work. He noted that there were many statements about the quantity of work done by the machines but never a word about its quality. He refuted the claim that one operator could do the work of three compositors (setting at 900 ems per hour) on the grounds: 'That there is no operator in Chicago who can turn out 2,700 nonpareil ems on an average. Of course I am aware of the fact that some employers claim that there are operators in their offices averaging nearly three thousand ems per hour, yet, notwithstanding this, I tell you that the 3,000 ems *are not there.*' He wrote: 'They will do well enough to grind out reading matter for the masses, and will undoubtedly result in the publication of a great amount of reading matter for sale at prices which the people will be able to pay.'

Critics considered hot metal mechanical composition inferior to hand setting because there was no kerning; particularly noticeable on the lower case letters f and *f* and their associated ligatures.

On the other hand it is only fair to note that some publications carried extremely enthusiastic reports; in particular *Life*, a British weekly journal, was certain that the Linotype would become popular and that the company would easily sell all the shares on the first launch. Thus printers were cautious while laymen were easily impressed.

## 2   *Contemporary reports of the impact of the Linotype*

In advertisements to launch The Linotype Company Limited the British Linotype Syndicate used sworn statements of setting speeds supplied by two US newspapers and Clephane's Philadelphia Linotype Company. These claimed that both men and women operators produced an average of about 3,000 ems (6,000 ens) per hour. The only person to suggest a much higher output was Miss J. Julia Camp, who had helped to demonstrate the second band machine at Baltimore in 1884. She 'composed from dictation, at one time at a rate of over 10,000 ems an hour' but it is not known how long the test lasted, whether someone read to her or whether she used a dictation machine of the sort shown in figure 40. Only an outstanding operator could have maintained such speeds over long periods, even on faster models of Linotype.

W. N. Haldeman, Proprietor and Publisher of the *Louisville Evening Times* and the *Louisville Daily Courier-Journal* declared that his four top operators had each set over 150,000 ems in a 55 hour week. W. B. Machen, previously mentioned in chapter 7, the second highest paid operator on the *Louisville Courier-Journal*, must have been one of the fastest operators of all time and seems to be the only one for whom records of output on Blower Linotypes in both the USA and the UK have survived. He set an American record on 17 February 1889 by producing 300,900 ems in a week, an average of 5,471 ems per hour. On 22 March 1890 the Linotype Company published details of his work at the Carlyle Press, 7 East Harding Street, London; he averaged nearly 8,000 ens corrected per hour in a 48-hour trial on a Blower machine and attended to all mechanical requirements. These figures of output by the same operator do not agree with the definition that one em was equal to two ens. If the working conditions in each country were similar then one American em would seem to be equivalent to about one and a half English ens.

Mr Dibblee, manager of the *Manchester Guardian*, found an even smaller ratio when he tried to resolve the apparent inconsistency during a trip to New York. He selected some matter from the *Boston Globe* which their foreman reckoned as 3,600 ems and the Manchester foreman reckoned to be 4,220 ens; much less than the expected 7,200 ens. This inconsistency might in part be explained by a 1903 report that American matrices were much broader than British ones. It was also claimed that American machines ran faster than British models but that would only speed up the casting process, it would make no difference to the speed of tapping the keyboard to assemble matrices. Therefore operating speeds in the USA were not really as fast as suggested by the figures.

Machen operated the improved Linotype (the first model of which had just arrived from America) when it was demonstrated for shareholders at the adjourned second ordinary general meeting of the Linotype Company at the offices of The Economic Printing and Publishing Company Limited, 9 Worship Street, Finsbury, London, EC, at noon on Friday 27 February 1891.

Clephane wrote to Mergenthaler on 1 January 1894 that Lee Reilly had approached the operating speeds predicted by J. Julia Camp. His letter ended with the statement: 'I never expected 10,000 ems an hour to be gotten *from copy*, but you see this figure has nearly been reached, and that in the ordinary course of business.'

He enclosed an affidavit that stated:

> Lee Reilly being duly sworn deposes and says, that he is a compositor on the Mergenthaler Linotype machine, (in the office of the *New York Tribune*) and that during the week ending December 20th, 1893 he set on said Mergenthaler machine, taking copy from the hook in the regular course of business, 400,200 ems of nonpareil matter, all of which appeared in print, the actual working time

being 46 hours and 10 minutes. Further, deponent says, that all of this matter except that of the last day was corrected by him within the time above named. Further, deponent says, that seventy-five per cent or upwards of the said matter was set solid, and that no hand work such as heads or leads was counted, and that no special preparation was made for doing such work.
Lee Reilly
Sworn to and subscribed before me this 29th day of December 1893.
G. T. Miatt, Notary Public for Kings County, Certif. filed in New York Co.

The above affidavit showed that Reilly averaged 8,668 ems per hour. The final sentence shows that the Americans and the British calculated the amount of setting by different rules.

Hand compositors went to work several hours before starting to set the newspaper to distribute the type from the previous issue. This was a boring chore during which my grandfather used to play chess with his particular friend – without a board! Although the composing room was never quiet, the noise consisted of human and manually produced sounds. By contrast there was a continuous mechanical background noise in a Linotype room from the clicking of driving belts, the clatter of 'mats' dropping into the assembler, the crash of lines being sent away for casting, and the tinkle of mats dropping into the magazine during distribution. There was little talking, because the operator had to remain alert; lapses in concentration could lead to accidents.

An American article claimed that:

> With the coming of the Linotype, out went the jovial, rollicking, liquor-drinking night compositor of the morning papers – never to return. Where one hundred of these formerly had filled a composing room, we now find twenty – perhaps thirty, papers are growing bigger – young men of a better than average calibre, sober, better educated, more accurate and capable. They are picked men – the average and the below average men are no longer wanted.

Before and after pictures of the composing room at the *Sheffield Daily Telegraph*, a prominent North of England newspaper, illustrate the speed and magnitude of the Linotype's impact on the trade. In 1892, Sir William Leng owner of that paper issued a pamphlet entitled, 'How we publish our papers,' which showed a large hand composing room and a Linotype room equipped with a battery of Blower Linotypes and a single Square Base machine, see figures 46 and 41. A further order of 12 improved Linotypes was soon to be delivered. Figure 47 shows that by 1901 the cases had been replaced by 27 Model 1 Linotypes and the Blower machines had vanished. Presumably some case hands were employed to set headings and make up pages. The value placed on these machines by newspaper proprietors can be inferred from the fact that, at that time, a Model 1 Linotype cost about ten times as much as a workman's cottage.

The introduction of composing machines that could do the work of five case hands altered the balance of labour and skills in the composing room, but the disruption was less drastic than had been feared, although it was traumatic for those displaced. In 1894 the *British and Colonial Printer and Stationer* forecast a possible displacement of 5% of the total labour force, the brunt of which would fall on the 'inexperts' and the 'grass hands'; a case of the survival of the fittest. The most intelligent men would go on to the machines, others would

DAILY AND EVENING COMPOSING ROOM.

Figure 46  The composing room at the *Sheffield Daily Telegraph*
*Source: Leng, How we publish our papers, 1892, p 14*

**THE LINOTYPE FOR NEWSPAPER WORK.**

ONE OF THE LARGEST PROVINCIAL INSTALLATIONS.

"SHEFFIELD DAILY TELEGRAPH."
(25 machines).

Figure 47   The composing room at the *Sheffield Daily Telegraph*
Source: *Linotype Notes, August, 1901*

become jobbers and the rest would find their occupation gone. The writer believed that the disruption would be both brief and minimised if employers and operatives would prohibit the introduction of further apprentices in all Society houses for three years.

It was noted that before the Linotype all printers suffered from a chronic shortage of type. The change from hand to machine composition was the biggest revolution in printing since the introduction of steam and the upheaval was probably greater. The centenary souvenir of the London Society of Compositors, contained the observation: 'The machine certainly did displace more than a few men, but the sequel is interesting . . . there was a sudden demand for more workers other than compositors . . . There was a big demand for proof readers . . . some of these old "pica-thumpers" became amongst the best proof readers the trade has ever known.'

The introduction of the Linotype upset the balance in associated parts of the printing industry. Typefounders who had relied on regular orders from newspapers for a new dress of type when their fonts wore out were seriously affected; printers who used to buy large quantities of type only ordered ingots, costing about one fifth of the price of a similar weight of type. Therefore the Linotype put some typefounders out of business and substantially reduced the income of others. In England, Joseph Lawrence told the 1896 annual general meeting that the Linotype Company was about to enter the field of the typefounding industry which was a large and important one. 'Typefounders had been attacking them in the past for the best of all reasons – that they had been injuring their business. But the typefounders in England had not the good sense to meet them in the spirit that they had in America. They [the Americans] did not imply that they were going to cease supplying Linotype machines. On the contrary, they were going to redouble their energies and make more of them.'

Many eminent typographers and printers welcomed the improved Linotype despite early misgivings about the reliability of the Blower machine and the quality of its output. John Southward, a severe critic of the original machine, changed his opinion of linecasting machines and endorsed the Square Base Linotype. By the end of 1895 he was advocating the use of Linotypes rather than typesetting machines that used movable foundry type. He remained a strong supporter of the Linotype and even dedicated a book to Joseph Lawrence, Head of the Linotype Company.

At first the compositors of the USA were furious about machines that did their work and there were threats of strikes but there were no reports of violence against either the machines or their operators in that country. When compositors found that they could earn higher wages as Linotype operators they accepted the change.

When the Linotype Company was advertised in 1889, some British papers consisted of a single folded sheet of four pages. *The Times*, with at least twelve pages, and sometimes twenty-four, was considered very large. The Linotype encouraged existing newspapers to expand and provided the means to introduce new ones, particularly in London's Fleet Street which produced national rather than regional newspapers. The Linotype was the key factor that made it possible for Alfred Harmsworth (later created Lord Northcliffe) to launch the London *Daily Mail* in 1896. At the end of the nineteenth century it was discovered that news and newspapers could be entertaining as well as factual. 'There was a new public to be served, no longer the upper and middle class elite; but the first products of universal compulsory education provided by the new elementary schools. People who could read, but did not want to be bored.'

The *Mail* made an immediate impact on Fleet Street; the *Sales and Wants Advertiser* reported that it was creating intense competition among London daily papers. The *Morning* increased its regular size from eight to ten pages, then the *Morning Leader* produced a 'mammoth' twenty-four pages for a halfpenny (one cent) and the big penny papers were printing enlarged issues.

Before the introduction of the improved web printing machine and the Linotype Composing machine it would have been impossible to produce the *Daily Mail* of eight large closely printed pages for a halfpenny. The report predicted that the labour market would benefit from larger penny papers and ended: 'It is a significant and interesting fact that, although there are nearly 200 Linotypes at work in London, there are today fewer compositors seeking work than there have been for years.'

Although the Linotype made it possible to produce the *Daily Mail* economically it needed Harmsworth to identify the gap in the market and exploit it. The *Mail* started a fashion in Linotype set halfpenny papers; it was followed by the *Daily Express* in 1900, the *Daily Illustrated Mirror* in 1903, the *Daily Sketch* in 1908 and the *Daily Herald* in 1912.

The Linotype was soon accepted, particularly at newspapers. Howe summed up the UK position in the following paragraph: 'Although the widespread introduction of composing machinery, and particularly the Linotype, caused the displacement of hand compositors, particularly in the provinces, the effect was only temporary. The constant expansion of the printing trade, both in and out of London, speedily absorbed any surplus of men and many more in addition.'

## 3  *Reminiscences of experience with early Linotypes*

Some fifty years after the first Linotype went into the *New York Tribune* people with Blower experience were asked for their memories of the old machines. They remembered people and places better than dates which is why these reminiscences have been separated from contemporary reports.

John T. Miller and Charles W. Letsch were two long surviving friends intimately associated with that first Linotype, see figure 48. Miller was the first regular Linotype operator on the *Tribune*, who became a proof reader after working as an operator for about 35 years. Letsch had been trained by Mergenthaler and helped him to install the first machine. He was a skilled machinist who installed Linotypes in many newspapers in the USA and in 1936 had been chief machinist with the *Philadelphia Record* for 44 years. He recalled being telegraphed at Brooklyn when the original machine went wrong one night, there being no telephones at the time, but he could not reach the *Tribune* because there was a fire and the police would not allow him through. To avoid a recurrence of the problem Whitelaw Reid gave him a reporter's badge which got him through fire lines and gave him free admission to nearly all the places of entertainment in the city.

Martin Q. Good who joined the *Tribune* on 15 June 1888 regretted the passing of 'the bitter personal newspaper battles when Whitelaw Reid edited the *Tribune*, Charles A. Dana the *Sun* and James Gordon Bennett the *New York Herald*.' He learned his trade in Slatington, PA, came to New York City at the age of 20 and spent many a hungry night on park benches before finding a job on the *Mail and Express*. The *Tribune* employed him because of his ability to read bad copy and put him in charge of a Blower Linotype. Although he knew little about the machine he was expected to take care of it; in those days the machinist often acted only as an advisor.

In October 1935, Alexander Gordon, a personal friend of the inventor, recalled working with him in 1892 and 1893 at the Baltimore factory. He knew him personally until his death and claimed that Mergenthaler was 'respected and beloved by all who had the privilege of knowing him.'

Harry P. Saunders was working at the *Derbyshire Times*, Chesterfield, in 1935 when he wrote about his experience of the Blower. He was an apprentice

# FACTS AND FANCIES

*Mergenthaler and Letsch, 1936*

*John T. Miller Meets an Old Friend*

*Miller and Letsch Fifty Years Ago*

Figure 48  Left: Miller and Letsch at Coney Island in July 1886
Above: Miller at the Blower keyboard
Right: Letsch with Mergenthaler bust
*Source: The Linotype News, August 1936*

This picture of John T. Miller, who operated the first Linotype, and Charles W. Letsch, who helped Ottmar Mergenthaler install that machine in the New York Tribune plant in July, 1886, was reproduced from an old tintype made at Coney Island fifty years ago, when Messrs. Miller and Letsch were enjoying a holiday away from the first old Blower at the *Scottish Leader* when the machines were introduced to Scotland. It was the first office to use Linotypes with American matrices to produce a daily newspaper in the UK. Other newspapers did not start to use Linotypes until British made matrices were available. He also operated the Blower at the shareholders' meeting at Worship Street, London, when Machen demonstrated the 'improved' Linotype but he erred over details. He put the date at 1889-90 instead of 1891 and described it as the first model of claw-base machine. At that time it must have been a Square Base Linotype. He recalled that the new model was known as the 'gravitation assembler' in those days. Mr Saunders final comment was: 'we had to operate in those days, not simply button-tap as they have to-day, when the machine runs so smoothly.'

Phil A. Chant operated a Blower at the Company's showroom in East Harding Street, Fetter Lane. It had been put at the disposal of the Carlyle Press after their premises had been destroyed by fire. The Company man in charge was a Mr Machin [just one of the variant spellings of the name.]

E. G. Leonard, one of the original Linotype operators on the *Sheffield Telegraph* identified his machine as being in the top right hand corner near the window, see figure 41, and recalled that Girod (one of Mergenthaler's colleagues) had erected the Square Base Linotype that appears on the left hand side of the illustration.

The January-February 1938 issue of the British *L&M News* included two important letters from the USA. The first came from C. W. Brookes who had been employed at the Economic Printing Company, a training and book printing branch of the British Linotype Company, in the early 1890s. About 1893–4, as he remembered it, he saw his first assembly belt (Square Base) Linotype installed 'in a glass encased room with a heavy Brussels carpet on the floor, and friend Machim [yet another spelling] operating it.' After graduating to the new machine with the 'copy cutter' keyboard he and others were sent to Cardiff to set the *Western Daily Mail*, whose premises had been burnt to the ground. They operated temporarily 'in a cow barn which still retained the odour of cows and horses.' He was unable to accept an offer to go to Belfast because his mother was seriously ill. This was fortunate in as far as 'the man who went was attacked in the street and man-handled as a menace to comps, which was not the only instance in the pioneering days.' However, the reaction to the Linotype was not usually so violent.

Brookes worked on London daily papers until 31 August 1896 when he sailed from Southampton to Australia on a two-year contract, and left Sydney for San Francisco in November 1898. He liked America so much that he finally settled down in Washington State.

The second letter written by Harry G. Leland from Baltimore was very important because he had been involved with the project from 1884. His letter corroborated details from several sources. Leland, who described himself as the first Linotype keyboard operator, was employed at Bank Lane by the National Typographic Company with Ottmar Mergenthaler, Julien P. Friez, Richard Berger, J. Henry Knoop, Charles W. Letsch, Ferdinand Wich, August Hoenisch and an errand boy. He correctly named people in the Washington group and the New York syndicate but thought that there were two Thorne typesetting machines at the *Tribune* instead of two Burr machines. [A Burr machine had been moved out to make room for the Linotype to come through the window.] He went to New York with the first machine, quoted the delivery date as June 1886 and said that the composing room foreman, W. P. Thompson, had paid the safe rigger $35 to guide the machine through the telegraph wires. When the machine was connected Reid came in and asked Leland if it was working all right, and being told that it was, gave him a message to set up. Leland decided to stay in Baltimore after Mergenthaler and Reid had differences. In July 1889, at Hine's request, he sailed for England to demonstrate the machine but recalled the London demonstration sites in the wrong order. After working in London he gave demonstrations at 42 W Nile Street, Glasgow and then at 45a Market Street, Manchester and returned to the USA in December 1889. He and Mergenthaler were very good friends and he assisted in the development of the improved Linotype whenever he could spare the time. Leland also kept in touch with Joseph Mackey, President of the Mergenthaler Linotype Company, who was making a history file, and he was still recording his memories in 1943.

## 4 *Nailing the myths*

There are two types of myth; first the genuine error in recalling details and second, deliberate distortion for effect; euphemistically called faction. The first form can also be the result of jumping to a conclusion; for example, The *Printer's Register*, 7 December 1891, page 8, col 2, reported: 'The inventor of the "Linotype" composing machine is neither a printer nor an engineer, but has been connected with bookbinding for many years. He is a partner in the Philadelphia firm of Oldach and Mergenthaler and is a German by birth.' There was such a firm, but the Mergenthaler partner was Adolph, not Ottmar!

In answer to a reader's question: 'When and where was the first Linotype installed?' the *L&M News*, July 1929, page 240, replied: 'On July 4, 1886, in the office of the *New York Tribune*.' It was the right venue, but the date was wrong. Unfortunately the exact date is not known; on 2 July 1886 Reid wrote to his wife that the machine was in and working beautifully, but 52 years later Leland said that it was in June – in any event it was before Sunday 4 July 1886!

When Mergenthaler resigned and the Mergenthaler Printing Company moved to Brooklyn, there were reports that he had moved his factory to Brooklyn – confusing the name of the company with the name of the inventor. However, Mergenthaler aided the misconception by allowing the firm to announce that he had resigned from the company rather than leave Baltimore.

It was widely believed, even within living memory, that Mergenthaler went mad designing the distribution system of the Linotype and died in penury in a lunatic asylum. I first heard the story in 1936 when I asked my father, who was demonstrating the Linotype at the Printing exhibition at Olympia in London, about the inventor. He also told me that the inventor was **Otto** Mergenthaler, a **Swiss** watch maker. The error in the forename was probably due to Mergenthaler signing himself '**Ott.** Mergenthaler'. In the 1906 article in the Canadian newspaper: 'How Otto Mergenthaler invented the Linotype' of the interview with Emma, she was reported to have called him Otto. Maybe the reporter misheard or possibly Emma actually called him Otto rather than Ottmar. Of course everyone knows that the Swiss make good watches and that the Germans make cuckoo clocks. It was pure fantasy to think that designing the distribution system could drive Mergenthaler mad, his method involved the logical application of binary arithmetic, although he probably did not realise the mathematical significance. Each matrix carried a unique combination of up to seven teeth that identified its letter. If the presence of a tooth is denoted by 1 and the absence of a tooth by 0, it can be shown that there are 127 possible arrangements with at least one tooth. The number 127 was reserved for extraneous sorts because a matrix with a full set of teeth could not fall until it reached the end of the distribution bar which had no retaining teeth. In 1936, John Miller recalled that Mergenthaler had made a detailed study of the distributor on the Burr machines at the *Tribune*, so it is unlikely that he would have found it too difficult to design. However, it had been reported that a patent examiner had gone mad trying to process the patent application for the Paige composing machine; this may have been attributed to Mergenthaler. Rumours of penury could have arisen because the inventor came to the USA as a poor youth and later was so short of money that he sold his shares for $60. Further, he complained about being swindled out of his rights. Sacrificing his holding in the National Machine Printing Company for $60 and the destruction of his plans in a fit of anger were examples of irrational behaviour that might have contributed to rumours of madness.

The second sort of myth came from deliberate distortion of the facts. Probably the most outrageous example is the report of how the Linotype was 'named' by Whitelaw Reid. This has become so widespread that it has been accepted without question for decades. The story usually puts Reid and Mergenthaler in the *Tribune* composing room just after the machine has been installed. Mergenthaler is demonstrating the machine to Reid who has just seen it for the first time. As the first slug is cast Reid says: 'Ottmar, you've done it – you've cast a line of type! Say that's a great name – let's call your machine a Linotype!'

There are many reasons for rejecting this story. Reid would never have called Mergenthaler 'Ottmar', it was not in the style of the 19th century; even Clephane was not so familiar. This false mode of address puts the rest of the report in doubt. Leland's letter to the *L&M News* said that he, not Mergenthaler, was at the keyboard when Reid entered. Further, the biography made it clear that Reid must have seen many slugs cast before he decided to invest in the project and it was unlikely that the inventor, who by all accounts was not an expert operator, would have given him a lesson on the Linotype. Reid had complained about the supply of matrices for months and test articles had been printed in his paper. The printing historian Bullen and Hutchins's biographer suggested that the machine was named by Stilson Hutchins. However, the article in the *Railway Press,* previously referred to in section 1, which must have been authorised by Hutchins, gave the credit to Mergenthaler. In any event the name Linotype did not supersede that of Mergenthaler for the machine for several years. It is possible that the Linotype name just evolved and if that were so it would be idle to speculate on its origin.

A worse example of crass fiction (although admittedly presented as fancy) was generated by the British Linotype Company in the *Linotype Matrix* of Spring 1939, see figure 49. The previous paragraphs show that the famous illustration by J. Coggeshall Wilson purporting to represent Mergenthaler and Reid at the *Tribune* when the first Blower Linotype was installed was complete fiction. Towards the end of this piece Mergenthaler is supposed to kiss a hot slug – he would not have done so twice – anyone who had picked up a slug

# The Birth of a Slug

Have you ever tried typographical jargon? Do you know your 'thins' and 'thicks' 'nuts' and 'muttons'? Can you 'impose' your 'forme' for the 'coffins' of the printing press with aplomb! What *is* a 'slug'?

*Here you see Whitelaw Reid and Ottmar Mergenthaler with the first Linotype machine in the 'New York Tribune' Office, July 1886*

IF YOU WERE an invisible spectator in any newspaper composing room, or any of those more enlightened up-to-date printing establishments which use the Linotype, you would not hear any solicitous enquiries, round about press time, as to the whereabouts of a certain quantity of rectangular metal bars, with raised characters in relief along one edge, the product of that remarkable machine the Linotype, rather would you hear an explosive 'where the b— h— are these *slugs*.'

The mumbo-jumbo of every profession is beloved by its own practitioners, but nowhere does it immediately tantalize by figurative association as in the printing trade. Compositors call loose type 'stamps'; they go in for 'make-up'; they 'dress' their made-up pages on a 'stone' with metal 'furniture'; the pages are then 'locked-up' in 'chases' with 'quoins,' and are then transferred as 'formes' to the 'coffins' of the printing machine.

More respectable than thieves' jargon, not so amusing as darting rhyming slang, printers technicalities had their day in the punning, pompous, latter half of the nineteenth century. Literally weak-minded compositors studded the minor literature of the craft with 'Songs of the Press,' in which they deepened the mystery by self-conscious double-entendres. But these effusions had gone out of fashion by the time the Linotype was in full development at the beginning of the twentieth century, so that it is difficult to trace the origin and increasing use of the term 'slug' as applied to the Linotype product. He was a sardonic humorist who first applied a slow, lazy, or sluggish connotation to these fast-flowing, easily handled units of a typesetting machine, whose speed and mechanical simplicity were quickly to revolutionize the production of the printed word. But we may be wronging that unknown genius, because he may have been startled by the ease and the rapidity of Linotype production into a comparison with the lead slugs used in early rifles, or the slugs still used by sons and fathers in air guns. On the other hand, such is the versatility of the English language, he (the aforesaid genius) may have been stunned, i.e., slogged or slugged (O.E.D: 1862, chiefly *North and U.S.*), struck, driven, thrown, etc., heavily or violently into a realization of the far-reaching significance of this new product.

We might even go further and reconstruct the occasion and the manner in which this unknown printer's devil coined another term for a world already overburdened with printers' technical expressions.

The date, shall we say, was July 1886. The occasion is the first test of the first machine set up in the printing office of the *New York Tribune*. Present: Whitelaw Reid, publisher of the *New York Tribune*; Ottmar Mergenthaler inventor of the Linotype; a Printer's Devil and Onlookers.

*DOOR OPENS AND CLOSES*
REID
'Morning, Mergenthaler. Everything ready?
MERGENTHALER
You see, Mr. Reid, it is quite simple. I take the piece of news copy and put it over here . . . so I can read it. Now I tap a letter . . . .
*KEY TAPPED, MATRIX SLIDES DOWN: SOUND OF SLIGHT ESCAPE OF AIR*
MERGENTHALER
Did you see it? That's the Matrix . . . the pattern of a letter. Whenever I push a key, the matrix of that letter drops down, just like that.
*AS ABOVE, CONTINUED SLOWLY*
MERGENTHALER
Now I push the next letter, and the next, and spaces at the ends of the words . . . and more letters . . . you see, Mr Reid, now there's a whole line of letter patterns. Now I turn this handle, and it makes all the letter patterns move over here to the left. . . .
*TO INDICATE ACTION DESCRIBED*
MERGENTHALER
To the metal pot, Watch now.
*LINE OF TYPE DROPS ON TRAY*
MERGENTHALER
You see, Mr. Reid, it's finished. It's type, Mr. Reid. A whole line of it.
REID
(*Slowly*) A line of type . . . A line of . . . Say, that's a good name for your machine! Linotype. Let's see you set some more type, Mr. Mergenthaler.
PRINTER'S DEVIL
Gee, what a slug (*sloig*) for the comps! (*fades*).
ONLOOKERS
(*Reeling*) Yeah! what a slug! what a slug! oi! oi! these poor comps. What a slug! (*fades*).
MERGENTHALER
(*Overhearing*) What you say, slug? What you mean slug? My beautiful logotype a slug! Like a slug from a gun, yes—but not slow like the other slug (*laughs quietly into his big black beard*). Gardeners, they say soot is good for slugs, but you my beautiful (*picking up metal slug*—*kisses bright metal*), your soot is the lamp-black in printers' ink.
REID
Here boy, boy!
BOY
Yes, sir!
REID
Tell the editor to send some of to-day's news copy here at once. We'll set it on our new machine. Our Linotype machine!
MERGENTHALER
Yes and print it from my beautiful slugs!
*ROAR OF THE PRESS: CRESCENDO: EXULTANT*

**Old Stuff!**

*Our grandfathers read Linotype-set newspapers—and to-day we still do the same. But it's not the same Linotype at work. Oh, no! Linotype keeps pace with new needs, new thoughts, new ways. Linotype is as modern as to-day's newspaper. Always ask for Linotype.*

Figure 49   A piece of fiction concocted by the Linotype Company
*Source: Linotype Matrix, Spring 1939, p 7*

straight from the machine would remember how hot it was. Reid was portrayed as portly; in fact he was very thin as shown by contemporary cartoonists, in figures 9 and 50. He was so concerned with his weight that he kept notes in his diary; for example in March 1876 he noted that he weighed $156\frac{7}{8}$ lb after a Turkish bath with a light towel and $177\frac{1}{2}$ lb clothed with a heavy overcoat. In August he was about three pounds lighter.

Fiction was presented as fact in two dramatised radio documentaries to celebrate 50 years of the Linotype: 'Salute to the Modern Newspaper,' broadcast by the NBC blue network on Monday 29 June 1936 and 'Ottmar Mergenthaler and the Invention of the Linotype,' presented on the WABC and CBS networks by Du Pont in 'The Cavalcade of America' series on Wednesday 1 December 1937 between 8.00 and 8.30 pm and repeated between midnight and 12.30 am.

The first broadcast was the winning script in a competition organised by the Mergenthaler Linotype Company. Eighty-five people entered and the prize of $500 cash and a trip to New York went to Charles A. Wright, a former newspaperman, who had become a journalism instructor in the School of Journalism at Temple University, Pennsylvania. The instructions for entrants included a potted history of the Linotype, most of which was correct, a schematic of how to write for radio and warning that the company reserved the right to modify the script. Wright's play was highly praised according to the *Linotype News* of August 1936, but contained little reference to Mergenthaler. There were errors in the script; Clephane addressed Mergenthaler as Ottmar and it was implied that Reid had not previously seen the machine but immediately gave it a name.

The Du Pont script concentrated on Mergenthaler and opened with his arrival at the Hahl shop in Washington. In the play Mergenthaler claims to be a good mechanic but does not mention his training as a watch maker until it slips out when he repairs a child's music box, an acceptable device to make a point, but there were many factual errors. Mergenthaler knew nothing about printing when he met Clephane and was doubtful about stereotyping when it was suggested. Someone must have decided to cast him as an absent minded professor who would go without food and had him ignore the lunch supplied by his wife Emma. In fact he had a keen appetite and, had he been married at that time, he would have wiped the plate clean. As in the previous broadcast Clephane called him Ottmar. The script writer omitted any mention of the band machines and the individual matrix machine appears as if by magic. The scene about installing the first machine was more realistic. The writer had obviously

LASHING HIMSELF INTO FEVER HEAT.

BLACK-LAW REID. "You are British Free-Traders, Dudes, Pharisees, Frauds, and Mugwumps—that's what you are!"

Figure 50  Another lampoon of Whitelaw Reid
*Source: Library of Congress*

followed up some of the stories in the *Linotype News* of August 1936 and included the ninth floor, the man on the Linotype guiding it through the telegraph wires, John Miller's brother working with Mergenthaler in Baltimore, and the 95 cents postage for sending slugs to New York, but the package contained more than the one slug mentioned in the broadcast! There was the usual fiction of Reid naming the machine, but at least he seemed to be aware of what was being delivered. However, the scriptwriter went over the top when Mergenthaler, an indifferent operator, was portrayed as writing and setting an article on the hoof, in order to give a brief outline of the working of the Linotype.

Samuel Fuller, Director, Producer and Author of the film 'Park Row', wanted to crown an outstanding newspaper career by making an epic. By undertaking all three major production rôles he avoided all constructive criticism, produced a curate's egg and apparently lost $200,000. He had started as a 12-year old copy boy on the *New York Journal*, was a reporter at 16 and a crime reporter and cartoonist a year later. He had worked on newspapers across the USA and always wanted his own newspaper but his treatment of the subject showed that although he knew the editorial side he had only a superficial grasp of the production side of the business.

In striving for realism he made his characters too intense and relied on dialogue rather than imagery. Fuller's hero, Phineas Mitchell, editor of the *Globe*, had a small office, but all the sets were so cramped that the film might have been made for early television. In trying to establish the period of action he took historical events of 1886 and distorted the time frame to fit his scenario which led to several gaffes – a pity because it obscured the point he was trying to make. He gave a fictitious version of Steve Brodie's famous jump off the Brooklyn Bridge (reported in both the USA and the UK) by placing it about three weeks before it happened. Fuller's editor decides to organise the subscription for the plinth for the Statue of Liberty some three months before it was erected. Actually Pulitzer of the *World* ran the subscription which was completed on 11 August 1885, nearly a year before the setting of this film. Like Pulitzer, Mitchell printed the name of every contributor to the fund even those giving only 1 cent, but Pulitzer used no collecting agents so that criminals who tried to prey on the good nature of the public could readily be identified; it was reckoned that under $10 was lost in fraud but there were some hoax contributions. Fuller put Mergenthaler, apparently unmarried, in a New York saloon with newspaper men and with no visible means of support, thus breaking faith with Mergenthaler's location and his cronies. The actor playing the inventor was a passable lookalike who smoked a large Bavarian pipe, but Mergenthaler was a cigar smoker. The editor, who was trying to start a new paper on a shoestring said, 'About that machine – we'll pick it up in the morning, have it ready, huh!' a clever trick that. What arrived was taller than any single magazine Linotype ever made. A mock-up of the front of a Blower, scaled up to cover the machine behind, was mounted in front of a later model Linotype, fitted with an automatic metal feed. People with no background in print may consider this as unnecessary hair-splitting but the scale should be right to make the image realistic. Fuller could have used a static model of the proper dimensions and cut in the delivery of a slug instead of showing a modern keyboard and galley, just as his 'hand compositor' was not really setting type. When Fuller's inventor described the machine to a rival newspaper owner he ascribed problems of the first band machine to the hot metal Linotype and had Mergenthaler say that he would be prepared to 'give' the machine to a reputable newspaper. Mergenthaler would never have given it away! Mergenthaler says that he calls his machine the 'Blower' because it used compressed air to move the matrices into the assembler, but it was not given that name until years after the introduction of later machines. Of course Mergenthaler did not get attacked and injured and never set a paper by himself. Although each minor distortion may have been trivial, the cumulative effect of so many is a travesty of fact that would misinform the layman and infuriate the informed. Fuller aimed for stark realism in his other work but unfortunately he let his passion for his subject cloud his judgement. A newspaper story with less crusading zeal and distortion of history would have made a much better picture.

## 5  *The last word*

When Whitelaw Reid died in 1912 he was US Ambassador to Britain and had been out of the newspaper world for some years. His obituary filled the first five pages of the *New York Tribune* on 16 December 1912 and hardly mentioned the Linotype which came in almost as an afterthought at the bottom of page 5, column 7: 'It was under his patronage and encouragement in the *Tribune* composing room that the linotype was made practical and was first put to practical use, and the company was organized which has since introduced it into the printing establishments of all the countries of the world.'

# TEN

## *The Whittakers and the last Blower Linotype*

The Whittaker family was probably unique in being involved in the sale and design of linecasting machines throughout the hot metal era. Peter Whittaker wrote that his grandfather, Matthew, had been apprenticed at the Machinery Trust, although there is no record of a registered company of that name before 1893. Whittaker's papers also referred to a book entitled 'The History of the Machinery Trust', but no copy has been located and it is not listed in the standard catalogues. Regardless of where he was trained Matthew Whittaker must have been an outstanding craftsman. Either late in 1888 or early in 1889, aged only 19 or 20 years at the time, he was sent to Baltimore to be trained by Ottmar Mergenthaler on the Blower Linotype and was said to have erected the first machine in the United Kingdom, but there are conflicting accounts of where this was. The tuition must have been organised by Hutchins and Peter wrote that his grandfather had been furious when he learned that the improved Linotype was being developed whilst he was being trained on the old model.

Matthew was a senior engineer with the Machinery Trust Limited and had several inventions to his credit. In particular, he invented the two-line or drop letter for small advertisements. *The Daily Telegraph*, the first Fleet Street newspaper to adopt the machine, refused to buy the Linotype until this feature was available. Matthew held several senior positions in the Linotype Company and was respected throughout the printing trade. His son Harold served an apprenticeship with the Linotype Company before the first world war and returned to the company after being invalided out of the army, but was dismissed when he refused an overseas assignment. This made conditions so difficult for Matthew that he resigned and, joined by Harold, started the business of M. H. Whittaker and Son. This enterprise became the main agent for Intertype in the North of England and also offered a service to Linotype users in the area.

Matthew so inspired young Peter with tales of the Blower Linotype that when he heard that there was a Blower Linotype at the Smithsonian Institution in Washington, DC he was anxious for details. By this time the business had converted to electronic technology and he asked his associate Chris Hall to visit the Smithsonian and examine the machine during a business trip to the USA. Chris found the exhibit and reported that it was just a load of junk.

On his next trip to the USA, Peter Whittaker went to the Smithsonian to see the state of the machine, and was appalled. He told the Curator of the printing exhibit that this was a unique machine and offered to refurbish it.

Sometime later he was surprised to be told that a package weighing 1.5 tons was waiting to be collected at Leeds airport. The electronic systems with which he worked were much lighter. When the package arrived at the firm it was found to be the Blower but several pieces were missing so it was necessary to make replacements and work out how the machine was originally set up. No regular service manuals had survived but the team were helped by patent specifications and Matthew's notes. The work was carried out by Whittaker employees, Gerry Procter and Geoff Poulter. A single Blower matrix, found among Matthew's effects, was used by L&M Ltd as a pattern to make a font for the machine. The keyboard layout had to be guessed because no marks remained on the keys. The lower case letters are probably placed correctly but an alphabet of small capitals has erroneously been put in the middle of the keyboard and the repeated letters and ligatures which were reported as being on the original machines are missing. This gleaming reconstruction is probably more splendid than any machine that came out of Mergenthaler's factory. It is believed that Peter Whittaker bore the whole cost of this work. Figure 51 shows him standing with Stan Nelson, a specialist at the Smithsonian, with the refurbished Blower in 1977.

The refurbished machine was presented to Mrs Nancy Armstrong, then American Ambassador to the Court of St James, who accepted it on behalf of the Smithsonian Institution, at a luncheon held under the chairmanship of the late Lord Goodman on 25 June 1976. The front of the menu was designed in the style of a nineteenth century Linotype advertisement for the Blower machine, shown much reduced as figure 52. This pastiche was made up of an advertisement for a Square Base Linotype from which the illustration had been removed, to be replaced by Bennett's woodcut of the Blower Linotype that appeared in *Engineering*, in 1889.

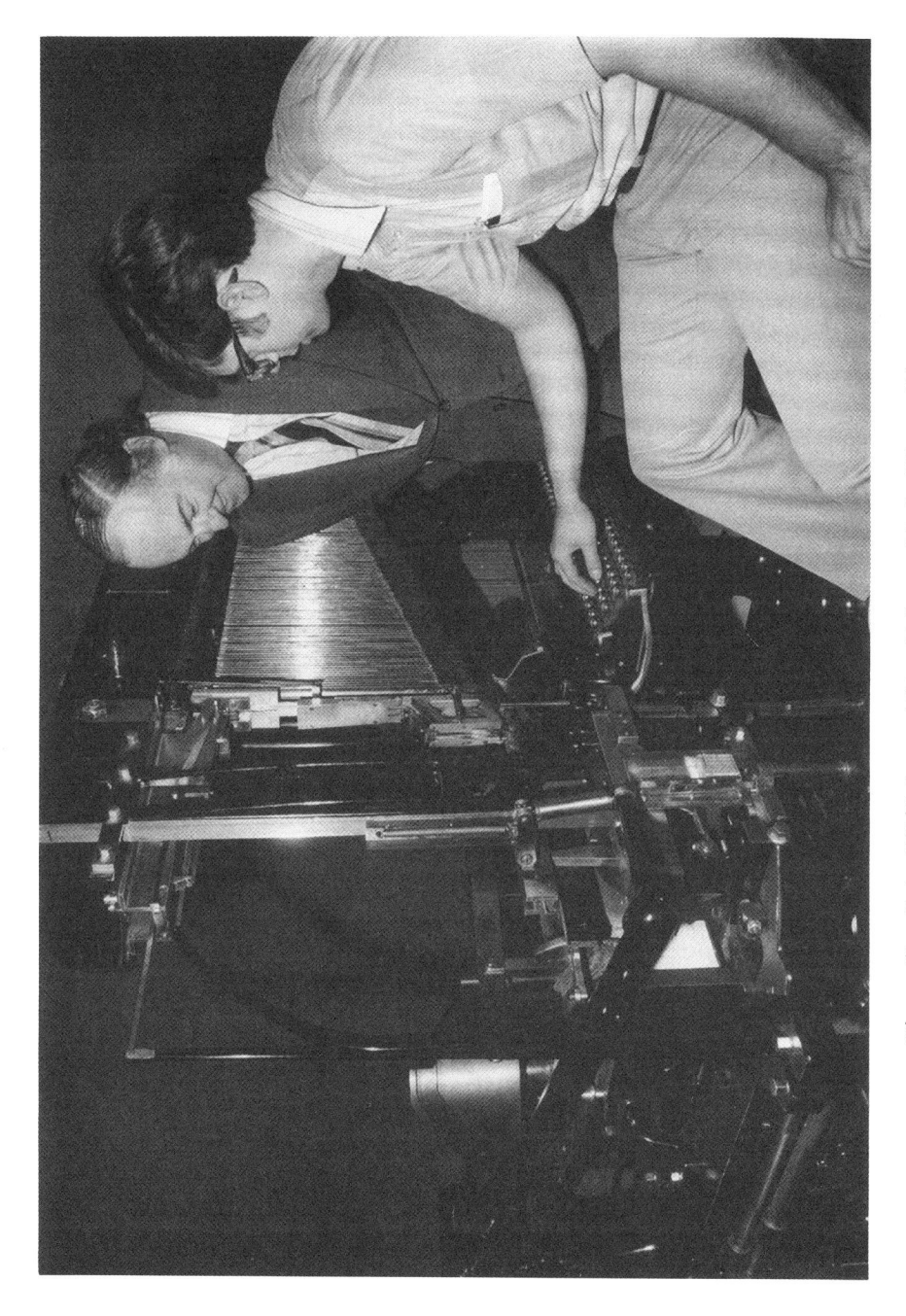

Figure 51  Peter Whittaker and Stan Nelson with the Blower Linotype
*Source: From an original photograph at the Smithsonian Institution*

## BLOWER COMPOSING MACHINE.

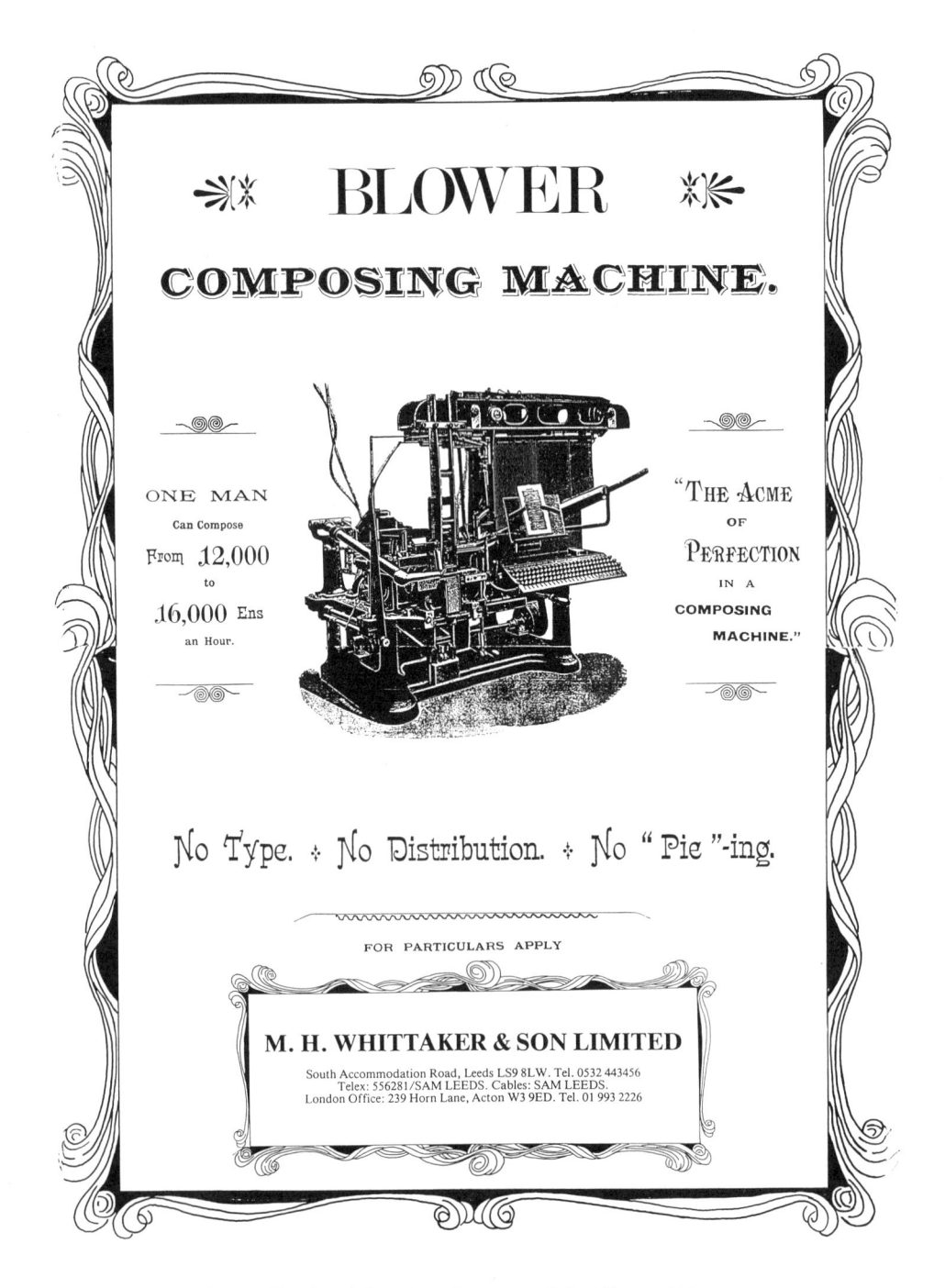

ONE MAN Can Compose From 12,000 to 16,000 Ens an Hour.

"THE ACME OF PERFECTION IN A COMPOSING MACHINE."

No Type. ∻ No Distribution. ∻ No "Pie"-ing.

FOR PARTICULARS APPLY

**M. H. WHITTAKER & SON LIMITED**

South Accommodation Road, Leeds LS9 8LW. Tel. 0532 443456
Telex: 556281/SAM LEEDS. Cables: SAM LEEDS.
London Office: 239 Horn Lane, Acton W3 9ED. Tel. 01 993 2226

Figure 52    Pastiche advertisement of the Blower Linotype
*Source: Menu at the presentation luncheon*

The inside of the menu gave information about the Blower machine and Peter inserted the name of his firm in the space left for a selling agent.

Before being returned to the United States the refurbished Blower was displayed in the United Kingdom at the Science Museum, in London, and by the BBC on the children's programme 'Blue Peter'.

Sometime later, Peter Whittaker formed a collection of some eleven linecasting machines, mainly Linotypes and Intertypes from the oldest to the most recent, although of course he did not have a Blower of his own. After his death in 1988 the collection and his papers were donated to the Museum of Science and Industry, Liverpool Road, Castlefield, Manchester, England, which is cataloguing his archives.

# APPENDICES

## A  *Selected Chronology*

This list shows the signposts in Ottmar Mergenthaler's life and significant events in the development and adoption of the Linotype in the United States and the United Kingdom. Any typesetting machines listed below are included because they were important during the period under consideration. In general, the dates given apply to the first known reference to the equipment, but like the Linotype, most of these devices were being developed continuously.

**1822** Dr William Church granted patent for the first known design of typesetting machine. Proposed an upright magazine containing sorts to be assembled by means of a keyboard and distribution through the melting pot.

**1854** Ottmar Mergenthaler born on 11 May at Hachtel near Bad Mergentheim; third child of village schoolmaster Johann Georg Mergenthaler (1820–1893) and his wife Rosina Ackermann (1828–1859).

**1859** Mergenthaler family moved to Ensingen. Ottmar's mother died.

**1861** Mergenthaler senior married Caroline Hahl, a loving and caring stepmother.

**1866** Clephane began his investigation of typewriters.

**1867** Mergenthaler repaired village clock at Ensingen.

**1868** After refusing to train as a teacher Ottmar started apprenticeship in watch and clock making with his stepmother's brother Louis Hahl.

**1871** Mergenthaler, an outstanding craftsman, paid as a journeyman during final year of apprenticeship.

**1872** Mergenthaler emigrated to the USA at the end of apprenticeship to avoid army service; sailed in SS Berlin, landed at Baltimore 26 October; went to Washington to work for step-cousin, August Hahl, building electrical appliances and models for inventors making applications for patent.

Burr typesetter announced. Used in several New York newspapers including the *New York Tribune* and the *New York World*.

| | |
|---|---|
| 1873 | The Remington Company started to manufacture typewriters commercially. |
| 1874 | Mergenthaler appointed foreman and deputy business manager by Hahl. |
| 1875 | Hahl workshop moved to Baltimore but maintained contacts with inventors. |
| 1876 | August 17, Charles T. Moore and James Ogilvie Clephane consulted August Hahl about a model of a transfer machine for producing a lithography master. Moore had been granted a patent in February of that year, but the model was unsatisfactory. Mergenthaler agreed to redesign it – his first experience of graphic arts. |
| 1877 | Moore machine completed, but output erratic. Clephane proposed stereotype, another process then unknown to Mergenthaler, to produce a matrix, one character at a time, for casting type in lines. |
| | National Machine Printing Company organised with capital of $28,000. |
| 1878 | Stereo machine completed at end of year. |
| | Wicks rotary typecaster announced; made 60,000 finished types per hour; vital for fast production of movable type for typesetting machines. |
| 1879 | Stereo machine gave erratic results; Mergenthaler left project; Clephane and associates continued research in a small workshop in Washington; Mergenthaler acted as part-time consultant to the group. |
| 1880 | Thorne typesetter announced; one of the most successful devices, later called the Unitype; used on some British provincial papers. |
| | The name of the Burr machine changed to Empire for British market and later was advertised in competition against the Linotype. |
| 1881 | Hahl made Mergenthaler a partner; Mergenthaler and Emma Lachenmayer married on 11 September. |
| 1882 | Clephane persuaded Washington attorney Lemon G. Hine to invest in the project. |
| 1883 | Mergenthaler dissolved partnership with Hahl; set up own workshop in Bank Lane, Baltimore; designed first band machine to produce stereo impressions in papier-mâché, one line at a time; Hine supported Mergenthaler by paying development costs of his invention. |
| | The National Typographic Company of West Virginia formed by Washington investors with capital of $1,000,000. |
| 1884 | Second band machine, the first direct line caster, demonstrated on 26 July at Bank Lane works in Baltimore by Ottmar Mergenthaler and Miss Julia Camp; Stilson Hutchins present; first time a slug was composed and cast on a Mergenthaler machine. |

## APPENDIX A: SELECTED CHRONOLOGY

Mergenthaler appointed factory manager with absolute control; new premises at 201 Camden Street; promised 10% royalty on cost of each machine and 1,000 $25 shares for producing a satisfactory machine.

Moore and Clephane gave up plans for a matrix making machine.

**1885** Stilson Hutchins organised demonstrations of second band machine and a banquet at the Chamberlain Hotel in February; eminent men, including Chester Arthur, then President of the USA, praised the inventor; Mergenthaler's speech.

During the year the Newspaper publishers' syndicate invested $273,024 (nearly $300,000) to gain control of the Linotype; many visited Camden Street works; E. M. Stone took over from L. G. Hine as chairman; Stilson Hutchins paid $200,000 finder's fee.

Mergenthaler produced first circulating matrix machine; prototype of all production models of Linotype; satisfactory trials in summer; deemed ready for production by October.

Benton punch cutting machine granted British patent.

J. W. Schuckers applied for patent on double wedge space band 49 days before Mergenthaler's application.

**1886** Mergenthaler Printing Company formed with capital of $1,000,000 to take charge of the business and share profits equally with National Typographic Company; Mergenthaler accepted 6% loan from Reid to secure allocation of shares; Linotype machines built in series; Mergenthaler started a matrix factory and invented 30 special machines for the production of matrices.

First Blower Linotype installed in *New York Tribune*, probably on 1 or 2 July; Whitelaw Reid reputed to have coined the name 'Linotype'; directors issued orders for 100 more machines despite Mergenthaler's misgivings.

**1887** *The Tribune Book of Open Air Sports* produced in January set by Linotype between news shifts; Mergenthaler modified keyboard on Blower Linotype; directors unanimously declared that machines were satisfactory and ordered 100 more on the basis of 12 delivered; directors started to interfere with Mergenthaler's management of the factory; Mergenthaler introduced bonus system to improve quality and speed up production.

Tolbert Lanston granted patents for first Monotype.

**1888** Whitelaw Reid's annual report mentioned early problems with the Linotype; Mergenthaler mistakenly interpreted this as an attack on the machine, contradicted him in open meeting and later resigned from the company.

Syndicate voted themselves exclusive use of Linotype at advantageous rates.

Stilson Hutchins went to England to sell sixty Blower Linotypes and the British rights after being offered generous commission.

Ottmar Mergenthaler started his own factory to produce Linotypes; Matthew Whittaker sent to Baltimore to learn about the Linotype from Mergenthaler; Improved Linotype (Square Base) developed with the aid of loan raised by Clephane.

Mergenthaler struck with pleurisy.

**1889** Reid's last report to shareholders before leaving company to become US Minister in Paris revealed quarrels over payment of royalties to Mergenthaler and commission to be paid to Stilson Hutchins for selling manufacturing rights to European companies; new board elected with Hine as President and General Manager; 12 machines sold to *Providence Journal*, first installation outside syndicate; Mergenthaler Linotype company started to use Benton-Waldo punch cutting equipment for making matrices.

Two machines set up in Stilson and Lee Hutchins's office to set Lawrence's Railway magazines.

Blower demonstrated in England; W. E. Gladstone commented favourably on the machine; Linotype Company Limited launched in Great Britain with £11,000 newspaper advertising campaign; first offer withdrawn and replaced by offer to pay Americans only in shares instead of shares and cash.

Mergenthaler constructed the Simplex (Model 1) that set the pattern for all future Linotypes.

Thorne started a determined advertising campaign as the first advertisements for the Linotype Company appeared.

**1890** First customer installations of Linotypes in Great Britain; British Linotype factory founded in Manchester, England.

Simplex machine exhibited in the Judge building in New York City; Mergenthaler Linotype Company founded in Brooklyn; Hine induced Mergenthaler to accept royalty of $50 per machine instead of the 10% stipulated in his 1884 contract with National Typographic Company.

Ottmar Mergenthaler awarded Cresson Medal by the Franklin Institute.

Rogers Typograph developed and installed in *New York World*.

**1891** Ottmar Mergenthaler awarded John Scott Medal by the City of Philadelphia.

Square Base Linotype demonstrated in Britain.

Linotype injunction against Typograph for infringement of patent issued 11 March; Rogers counterclaimed for infringement of patent on Schuckers sliding wedge space band.

**1892** First British built Square Base Linotypes installed at *Newcastle Evening Chronicle*; models produced in the British factory had different features from the American machines.

## APPENDIX A: SELECTED CHRONOLOGY

Wilbur Scudder, superintendent of the Linotype factory produced the Monoline with help from Hine and resigned in June.

Mergenthaler travelled to Germany to meet his father again.

**1893** Scales for machine setting agreed by the London Society of Compositors (LSC); Linotypes installed in London newspapers.

British Linotype company introduced 2-line letter before Mergenthaler.

Linotype training schools established in London and Manchester.

**1894** In January Mergenthaler devised the step justifier to circumvent the problems arising from the Schuckers case; made 225 machines of this design in Baltimore; Mergenthaler Linotype Company bought out Rogers Typograph to end stalemate over litigation; Rogers appointed consulting engineer to Mergenthaler company.

In October, the first Linotype machine was installed on the European continent. A British model to set *De Neederlandsche Financier* in Amsterdam. First edition postponed for five days by protest strike of the newspaper employees.

Mergenthaler's associate, C. A. Albrecht, and newspaper publisher William Mayer with son Jacques arrived in Germany in November. At the same time the first Linotype in Germany was shown in a small Berlin shop.

Ottmar Mergenthaler struck with tuberculosis and went to the southern part of the United States.

**1895** Star-based Model 1 announced in Great Britain; first machine delivered to the *Wakefield Chronicle*.

**1896** New favourable scales agreed for Linotype setting in London.

The Mergenthaler Setzmaschinenfabrik GmbH founded in Berlin.

**1897** Mergenthaler started to write his memoirs in Deming, New Mexico. All his belongings including books and manuscripts destroyed in prairie fire. Dictated a shorter biography to Otto Schoenrich.

Linotype Company works established at Altrincham.

**1898** Mergenthaler's biography printed at his own expense for private distribution. [Reid employed agents to acquire and destroy copies of the biography. The agents were reported to have offered up to $75 per copy.]

**1899** Ottmar Mergenthaler died in Baltimore, 28 October. Buried in the Loudon Park Cemetery.

# B  *Glossary*

**agate**, traditional name for American type size now called 5½ point.

**assembler**, or assembly box, the point where matrices and space bands are held while the Linotype operator is composing a line.

**bourgeois**, [ber-joyce] traditional name for type size now called 9 point.

**brevier**, [br'veer] traditional name for type size now called 8 point.

**burr**, or hair-line[2], thin lines of metal that appear between characters if the matrix walls become damaged.

**case hand**, a hand compositor.

**case**, a shallow tray divided into compartments for storing movable type; see frame, lower case and upper case.

**composing stick**, a hand held frame used by compositors to set movable type.

**composing**, setting matter in type or by matrices with spacing between words.

**compositor**, a person who sets type, usually by hand from a case, but also applied to an operator of a typesetting machine.

**currency**, in the 19th century £1 was worth US $5. There were 20 shillings in £1 and 12 pence (pennies) in a shilling. A halfpenny (haypny) was about 1 cent.
A guinea was a pound and a shilling.

**distribution**, replacing type or matrices and spaces into case or magazine.

**distributor**, a mechanism for distributing matrices into the magazine after a line has been cast on a Linotype.

**dress**, (of a periodical) a term to describe the appearance and style of a publication, particularly its typography and layout.

**elevator**, a device for raising or lowering matrices and space bands in a Linotype, see first elevator and second elevator.

**em**, originally the width of a letter m, now applied to the width of the body size of the type in use. The 12 point em, approximately one sixth of an inch, is also used as a unit to measure line width. In the USA a compositor's output was measured in thousands of ems.

**en**, half of an em. In the UK a compositor's output was measured in thousands of ens.

**etaoin/shrdlu**, the lower case characters on the two leftmost columns of English language Linotype keyboards taken from top to bottom.

**extraneous sort**, a character that is not on the Linotype keyboard. It is placed in the assembler by hand, has a full set of teeth, and drops down a chute for extraneous sorts at the end of the distributor.

**face**, a particular design of type.

**first elevator**, a sliding device that takes a line of matrices and space bands from the assembler down to be cast and then up to the level of the second elevator.

**flong**, papier-mâché sheets used for making stereotype matrices.

**font**, the complete set of characters that makes up a particular style and size of type.

**foundry type**, movable type supplied by a foundry.

**frame**, a stand for holding type cases from which a compositor works.

**galley**, a flat tray for holding composed type. In the Linotype, a device for holding lines of type as they are ejected from the machine.

**grass hands**, casual workers.

**hair-line**[1], thin strokes of a type character.

**hair-line**[2], or burr, thin lines of metal that appear between characters if the matrix walls become damaged.

**justification**, spacing words in a line equally to fill the line and present a smooth right hand edge to the text.

**kern**, the part of type that overhangs an adjacent character, particularly for characters such as italic *f*.

**ligature**, a logotype, usually of two characters.

**liner**, metal pieces inserted in a Linotype mould to set the body and measure of a line.

**Linotype**, a linecasting machine made by the Linotype Company.

**linotype**, a type bar, or 'line of type'.

**locking-up**, the process of securing a page of type firmly for printing.

**logotype**, two or more characters cast together on the same body.

**long primer**, [long primmer] traditional name for type size now called 10 point.

**lower case**, a synonym for small letters, or the case which held them and was laid almost horizontally at the front of the frame.

**machinist**, an American term for a skilled engineer, who might invent, construct, maintain or repair machinery; one skilled in the use of machine tools.

**magazine**, the container for Linotype matrices.

**matrix**, a recessed mould from which to cast a letterpress character.

**minion**, traditional name for type size now called 7 point.

**movable**, separate pieces of type for assembly by a case hand.

**nonpareil**, [non-prull] traditional name for the type size now called 6 point.

**patrix**, a punch used for making a matrix. This term is a correlative term to matrix, but etymologically incorrect.

**pearl**, traditional name for type size now called 4¾ point.

**pica**, [piker] traditional name for type size now called 12 point.

**pie font**, a set of extraneous sorts.

**pie**, movable type that has been jumbled.

**point**, a unit for measuring type size 0.013837 inch, approximately 1/72 inch.

**punch**, a piece of steel engraved with a type character used for stamping a matrix.

**QWERTY**, the first six letters taken from the left of the third row of keys on a standard English language typewriter.

**ruby**, traditional name for type size now called 5¼ point.

**second elevator**, a device that takes the matrices that have been raised by the first elevator and swings them up to the distributor.

**small pica**, [small piker] traditional name for type size now called 11 point.

**sort**, a character, either type or a matrix.

**space band**, a sliding double wedge device for justifying lines of type.

**splash**, an English term for the result of molten metal being forced between either the metal pot and the mould or the mould and the line. It can occur if the machine tries to cast a line when the metal pot is not tight against the mould, the justification mechanism is not working properly or if the operator sets two space bands together. It is usually due to bad maintenance; see squirt.

**squirt**, an American term for a splash.

**stereotype**, a letterpress plate cast of a papier-mâché flong or plaster mould of letterpress setting.

**swift**, a fast compositor or Linotype operator in North America.

**the trade**, referred to in context means the printing trade.

**type bar**, a linotype.

**type sizes**, originally specified by name, now identified by point size; see also: agate, bourgeois, brevier, long primer, minion, nonpareil, pearl, pica, ruby, small pica.

**upper case**, a synonym for capital letters, or the case which held them and was mounted nearly vertically behind the lower case.

**whip**, a fast compositor or Linotype operator in the UK.

# C  Biographical notes

**William Blades** (1824–1890), British printing historian and master printer was highly respected throughout the trade and able to influence printing practice. He was a strong supporter of trades unionism. His critical letter about the Blower Linotype may have influenced Southward's comments about the machine.

**Jacob Bright**, MP (1821–1899) was younger brother of the more famous John Bright. They were sometimes confused in press reports because of the common initial. Jacob Bright was head of the syndicate that promoted the Linotype in the United Kingdom and named as party of the first part in the initial papers setting up the Linotype Company in the UK. He probably invited William Gladstone, leader of the Liberal party, to attend the demonstration of the machine and make his widely quoted remarks about the capabilities of the machine. Gladstone may have resented this blatant commercialism; in any event Bright's connection with the Linotype Company was not mentioned in any subsequent biography even though he was chairman of the company on a salary of £1,000 for about six years until he resigned through ill health in 1895.

**James Ogilvie Clephane** (1842–1910), a brilliant shorthand writer who was private secretary to William H. Seward, Secretary of State during the American Civil War, became a leading court reporter. He qualified as a lawyer and was admitted to the Bar of the Supreme Court of the District of Columbia but did not have a regular practice. He was the first serious user of the typewriter, testing several prototypes to destruction, which led to the production of reliable models by the Remington Company. His primary aim was to produce better law reports and he proposed that lithography masters be made by typewriter. When this technique failed he suggested other methods that ultimately led to the invention of the Linotype by Ottmar Mergenthaler. He put all his resources into his early speculations and could not afford to invest in the Mergenthaler Printing Company.

**Philip Tell Dodge** (1851–1931), engineer, inventor and patent attorney, processed most of Ottmar Mergenthaler's patents and contributed many significant improvements to the Linotype. He concentrated on improving the quality of the Linotype printing surface and his invention of the two-letter matrix was described as the most important advance since the introduction of the improved Linotype. For many years he headed both the American Mergenthaler Linotype Company and the British Linotype Company. As a young man Dodge was reprimanded for acting without authority when he purchased the patents for the shift mechanism on the typewriter on behalf of the Remington company. On the instructions of Whitelaw Reid he went to the Northwestern Type Foundry in Milwaukee to investigate the Benton-Waldo punch (patented in 1885) and arranged a leasing agreement which led to the production of high quality matrices.

Patent records show that Philip Tell Dodge, working with his father, had processed patent applications for Clephane and others six years before he first handled a patent for Mergenthaler. Dodge prepared Mergenthaler's first patent which was granted on 26 August 1884. With only four exceptions, Dodge acted as patent attorney for Mergenthaler's applications from 1884, even after he became President of the Mergenthaler Linotype Company. Some of these records show that Mergenthaler had applied for patents on machines involving new concepts before trying to convince his backers to support them.

**Stilson Hutchins** (1838–1912) founder of the *Washington Post* in 1877, was the first person to realise the potential value of the Linotype in newspaper production. He attended the first demonstration at Mergenthaler's Bank Lane workshop and organised demonstrations and a banquet at the Chamberlain Hotel. Then he started to promote the machine in earnest and gained the interest of a syndicate of wealthy newspaper men with the aim of taking over the project. Hutchins was a handsome, swash-buckling, ruthless, go-getter, with an honest open countenance who received large fees both for finding the Linotype for the Syndicate and for selling the manufacturing rights to the British Syndicate that was organising the Linotype Company Limited in the United Kingdom. He set up offices in London to advertise and demonstrate the Linotype before the company was launched in the United Kingdom. This man, who acquired the title 'Hon' from being elected to a state legislature, was in many ways a thoroughly reprehensible character, yet, without his greed and drive the Linotype may never have got beyond the experimental stage. Although wealthy, he used to bum pennies for tramcar fare and reckoned that 'being a little short' had saved him over $100,000 during his lifetime. He was married so many times that it became a joke. He never paid the alimony to which his first wife was entitled and the payments were being disputed between their estates years after both were dead. He gave Joseph Pulitzer his first job writing in English and served with him in local government. John Philip Sousa was commissioned, by the newspaper, to write the *Washington Post* march, presumably on the orders of Stilson Hutchins.

**Joseph Lawrence,** MP (1848–1919) was proprietor of the *Railway Press* and the *Railway Herald* with offices at Southampton Buildings off Chancery Lane in London. He rented offices to Stilson Hutchins and his son Lee to start demonstrations of the Linotype for launching the company in Great Britain. Followed Jacob Bright as Chairman of the Linotype Company Limited and was also Chairman of the Machinery Trust. He became Chairman of Linotype and Machinery when the companies merged. He set up the original negotiations with the Americans and was responsible for getting the price reduced when it was realised that the Americans had sold obsolescent machines to the British company.

# APPENDIX C: BIOGRAPHICAL NOTES

**Friedrich (Fritz) Mergenthaler** (1868–1938), Ottmar's half-brother was the only close relative to follow him to the United States. He helped to install Linotypes in France and Belgium (to protect patent rights) before coming to America in 1889 to work at Locust Point. In 1894 the brothers argued over remuneration and Fritz left. The breach was never healed, Fritz was not mentioned in Ottmar's will or his biography. After leaving the Mergenthaler works Fritz worked as a Linotype installer and repairman before becoming a machinist at the *Baltimore Sun*.

**Ottmar Mergenthaler** (1854–1899) one of the outstanding inventors of all time and principal inventor of the Linotype was a brilliant craftsman who refused to become a teacher as his father wished. He was trained as a watch and clock maker by his stepmother's brother, Louis Hahl. In 1872, at the end of his apprenticeship, he emigrated to the USA, partly to avoid conscription into the army and went to work for his step-cousin, August Hahl, where he made signalling equipment and models for inventors applying for patents. He had no experience of printing before James O. Clephane came to the Hahl shop for assistance and until 1883 was in turn an employee and partner of Hahl. After 1883 he assumed leadership of the development work that led to the Linotype.

When the Typograph case arose he invented a form of single wedge justification that was fitted to several Baltimore constructed Linotypes, but was abandoned when the Linotype Company bought the rights in the Schuckers patent for $416,000.

**Charles T. Moore** (1847–1910) an inventor from White Springs, West Virginia, who was approached by J. O. Clephane to produce a writing machine from which a lithography master could be produced. When his invention did not work properly he and Clephane sought the advice of August Hahl and Mergenthaler became involved.

**The Newspaper Syndicate** was a group of powerful men who took over management of the National Typographic Company and inaugurated the Mergenthaler Printing Company to finance the production of Linotypes. It was led by Whitelaw Reid of the *New York Tribune* and included W. N. Haldeman of the *Louisville Courier-Journal*, Victor Lawson and Melville Stone of the *Chicago News,* W. H. Rand of Rand McNally & Co, Stilson Hutchins of the Washington Post and William Henry Smith.

**Whitelaw Reid** (1837–1912) was an eminent public figure as well as a prominent journalist. He was a university graduate who, under the pen name Agate, contributed some outstanding reports during the American Civil War for the *Cincinnati Gazette*. He was appointed (c 1863) librarian to the Library of Congress. Horace Greeley brought him on to the *New York Tribune* as Managing Editor and Reid took over the paper after Greeley's death. Between 1872 and 1905, he was editor-in-chief and chief proprietor of the *New York Tribune*, much of the time as an absentee editor. He left the paper when he was appointed United States Ambassador to the Court of St James. The new Tribune building, erected in 1875, was New York's first skyscraper. Somewhere along the line he seems to have developed into a bigot who threw the union out of his newspaper over a strike in 1877. The union was not reinstated until 1892. Contemporary cartoonists lampooned him for his extreme views.

In 1881, Reid married Elisabeth, the daughter of Darius Ogden Mills, an extremely wealthy Californian banker. Both Reid and Mills had large holdings of stock in the Mergenthaler Printing Company. Some idea of the wealth behind him is shown by the fact that his father-in-law spent $60,000 on decorating the library in his New York mansion and had a country estate, 'Ophir Farm,' near White Plains, where the opulent life style was said to rival that of the crowned heads of Europe.

Mergenthaler thought that Reid had been voted off the board for inefficiency but he had to resign when he was appointed US Minister to Paris. Reid returned to the USA in 1892 and was losing Republican vice-presidential candidate in the elections that year. He held many prominent positions and when Mergenthaler published his privately printed biography in 1898, was so disturbed at the threat to his image that he hired agents to buy every copy possible at prices up to $75 each, just to destroy them. Whitelaw Reid died in office and was so highly regarded by the British that they held a memorial service for him in Westminster Abbey and his body was returned to the USA in a British battleship. On the surface a cultured gentleman he was a ruthless business man who would brook no interference. The *New York Tribune* printed a five-page obituary in which there was only a short reference to the Linotype in the final paragraph.

**John R. Rogers** (1856–1934) inventor of the Typograph graduated in 1875 from Oberlin College, Ohio where his courses prepared him for teaching Greek and Physics and for Mechanical Engineering. He worked on bridge construction for the Des Moines & Northwestern Railway and became division superintendent in construction. Later he was engineer in charge of all bridge work for Wisconsin Central Railway. No doubt stimulated by youthful experiences as a hand compositor at Berea he formulated a design for an impression type composing system when he was superintendent of schools at Lorain, Ohio and after 1888 concentrated entirely on type casting and setting machines. He invented the Typograph which was judged to have infringed Linotype patents and was suppressed; but, in its turn, the Linotype space band was judged to have infringed the patent of the Schuckers space band, which had been acquired by Typograph. The impasse was broken when the Mergenthaler Linotype Company bought out the Typograph for $416,000 and retained Rogers as a consultant, a post he held until his death.

**Jacobs William Schuckers** (1831–1901), the son of a German father and Irish mother, was one of the few people involved with the Linotype who was trained as a printer. He left school in 1846 and worked at the *Wooster Republican* and then at the Cleveland, Ohio *Leader*. He became a clerk in the United States Treasury in Washington and in 1861 became private secretary to the Hon Salmon P. Chase who had been appointed Secretary of the Treasury. He went to Albany, New York, when Chase became Chief Justice of the United States Supreme Court, and studied law, but did not complete his studies. He wrote a *Life of Chase*, and for years contributed to the *Sun* and other New York papers, as well as to the press of Philadelphia. He speculated in real estate until the panic of 1873 swept away his little fortune. Then he began to devise a typesetting machine, culminating in a model with a typewriter keyboard that incorporated his double wedge spacer. In 1901 he was secretary for the New Jersey Commission at the Pan-American Exhibition in Buffalo.

**Wilbur S. Scudder** (1859–1932) was a tool maker who went to work for Mergenthaler in 1887. He replaced Charles H. Davids as superintendent of the Brooklyn factory in 1889. Scudder developed the Monoline in 1892. This had many features in common with the second band machine and was suppressed in the USA and the UK. It was probably an attempt to find sufficient originality from the previous patents to be awarded a patent in its own right. Later he worked for the Intertype Company.

**John Southward** (1840–1902) printer, author and technical editor was an acknowledged expert in his field. Amongst other things he contributed the article on Typography in the Encyclopaedia Britannica and catalogued the Blades Collection for the St Bride Printing Library. He criticised the Blower Linotype adversely, was threatened with legal proceedings by the Linotype Company and publicly challenged them to take him to court. He praised the next model, the square base, warmly and from then on was an ardent supporter of the machine.

**The Washington Group** was led by Lemon G. Hine and James Clephane. Among its members were Lewis Clephane, J. H. Crossman, Abner Greenleaf, Frank Hume, Stilson Hutchins, Kurtz Johnson, E. V. Murphy, Maurice Pechin and Frederick Warburton.

**Matthew H. Whittaker** (1869–1955) described by his grandson Peter as a senior engineer with the Machinery Trust, although aged only nineteen at the time, was sent to Baltimore to be trained by Mergenthaler. He is reputed to have installed the first Linotype in England in 1889, and was said to have been furious when he realised that the Americans had concealed the existence of the improved Linotype. He contributed several inventions to improve the machine and patented a two line letter (drop letter for advertisements) before Mergenthaler produced one in the USA. (This was one of the requirements for *The Daily Telegraph* to adopt the Linotype for setting their paper.) He held several senior positions in the Linotype Company and was respected in the printing trade, but resigned when his son Harold, who had served his apprenticeship with Linotype before the first world war, was dismissed because he refused to take an overseas assignment. Matthew and Harold started the business of M. H. Whittaker and Son which became the main agent for Intertype in the North of England and also offered a service to Linotype users in the area.

**Peter Whittaker** (1926–1988) grandson of Matthew was fired with enthusiasm for the Blower Linotype by his grandfather's reminiscences. He went to the Smithsonian Institution in Washington, DC to see the only surviving Blower Linotype; later, his company refurbished the machine.

# Bibliography

## Archival material:

Agreement between Ottmar Mergenthaler and The National Typographic Company, dated 13 November 1884.

Henderson A. L, *Baltimore Street Addresses of Ottmar Mergenthaler*, unpublished report from the files of the Commission for Historical and Architectural Preservation of the City of Baltimore. Addresses obtained from Baltimore City Directories.

Mergenthaler Printing Company, *Annual Report*, 1888.

Mergenthaler Printing Company, *Annual Report*, 1889. [Manuscript .]

Patent Specifications – British in hard copy, American on microfilm at the Patent Office, Chancery Lane, London, England:

- US patent 201436, Preparing transfer sheets or matrices for printing, granted to Charles T. Moore, 19 March 1878.
- US patent 304272, Matrix making machine, granted to O. Mergenthaler, 26 August 1884. [Mergenthaler's first patent, prepared by P. T. Dodge.]
- US patent 317828, Machine for producing printing bars, granted to O. Mergenthaler, 12 May 1885. [Blower prototype.]
- US patent 378798, Machine for producing type bars, granted to O. Mergenthaler, 28 February 1888. [Blower Linotype.]
- US patent 436532, Machine for producing linotypes, type-matrices, &c, granted to O. Mergenthaler, 16 September 1890. [Square Base Linotype, application filed 11 November 1889.]
- US patent 547633, Double-letter matrix, granted to P. T. Dodge, 18 October 1895. [Application filed on 25 May 1892 but invention not announced for sale until 1898, for roman with italic and small caps.]
- British patent 10257, Forming two-line capitals on type-slugs, granted to Matthew Whittaker, 1894. [But existence noted in 1893, in 'The Linotype: its history, construction and operation', p 5.]
- British patent 11615, Mechanism for metal-feeding to pot, granted to W. H. Lock, J. Place and W. J. Lewis, 1898.
- British patent 21542, Keeping molten metal at constant level in pot, granted to P. T. Dodge, 1901.

## Dissertations:

Goble G. C, *The Obituary of a Machine: The Rise and Fall of Ottmar Mergenthaler's Linotype at US Newspapers*, PhD Thesis, Indiana University, November 1984.
Kahan B. C, *The Linotype in the United Kingdom 1889–1912*, PhD Thesis, Reading University, England, September 1994.

## Directories and dictionaries:

*Dictionary of American Biography*, New York, Charles Scribner's Sons, 1981.
*Dictionary of Business Biography*, London, Butterworths, 1985.
Law Society lists for the years 1888–1891.
Philadelphia City Directories for the years 1884, 1885 and 1886.
*Sell's Dictionary of the World's Press*, 1892–1895.
Who Was Who in America, Chicago, Marquis Who's Who, vol 1, 1897–1942, fifth printing, 1962.
Who Was Who, 1897–1916, London, A & C Black, 1920.

## Books and articles:

Annenberg M. (Compiler), 'A Typographic Journey through the *Inland Printer* 1883–1900', Baltimore, MD, Maran Press, 1977.
Arnold E, 'Shocking Fate of the Unique Mergenthaler Collection', *Type and Press*, c 1988, #48, USA.
Ashworth J, *Operation and Mechanism of the Linotype and Intertype*, London, 2 volumes, Staples Press Limited, 1955.
'Blower Linotype Operators Speak for Themselves', *L&M News*, November 1937, p 29.
Bullen H. L, an important series of articles in the *Inland Printer*:
– 'The Transition from Hand-Set to Machine-Set Composition', Chicago, *Inland Printer*, January 1924;
– 'Origin and Development of the Linotype Machine', parts I and II, Chicago, *Inland Printer*, February-March, 1924, pp 769–771 and 936–8;
– 'The Typograph and the Monoline Machines', Chicago, *Inland Printer*, April 1924, pp 65–67;
– 'Origin and Development of the Lanston Monotype Composing Machine', Chicago, *Inland Printer*, May 1924, pp 228–32;
– 'The Effect of the Composing Machines Upon the Typefounding Industry', Chicago, *Inland Printer*, July 1924, pp 595–7.
Clephane J. O, *History of the Typewriter and the Mergenthaler Linotype Machine*, no imprint, undated, but including a copy of a letter from Mergenthaler to Clephane, dated 18 July 1894.

Clephane J. O, *History of the Typewriter, the Mergenthaler Matrix Setting and Casting Machine, the Graphophone, the Improved Phonograph, the new Fowler and Henkle Printing Press and What Led to Their Development*, no imprint, December 1889.

Dawson J, *An Editor's Experiences with the Linotype*, typescript starting on p 47 of one of Southward's notebooks at the St Bride Printing Library.

Dreier T, *The Power of Print – and Men*; New York, The Mergenthaler Linotype Company, 1936.

Eckman J, *The Heritage of the Printer*, Philadelphia, PA, North American Publishing Co, 1965.

Gallagher E. J, *Stilson Hutchins 1838–1912*, Laconia, NH, Citizen Publishing Co, 1965.

Goble G. C, *Rogers's Typograph Versus Mergenthaler's Linotype: The Push and Shove of Patents and Priority in the 1890s*, 35 Printing History, Vol XVIII, Number 1, The Journal of the American Printing History Association, December 1997, pp 26–44.

Hall H. (Editor), *The Tribune Book of Open Air Sports*, New York, The Tribune Association, 1887.

Howe E, *The Trade*, London, privately printed by Walter Hutchinson, 1943.

Huss R. E, *Dr Church's 'Hoax'*, Lancaster, PA, Graphic Arts, Inc, 1976.

Huss R. E, *The Development of Printers' Mechanical Typesetting Methods, 1822–1925*, Charlottesville, VA, University Press of Virginia, 1973.

Huss R. E, *The Printer's Composition Matrix*, New Castle, DE, Oak Knoll Books, 1985.

Iles G, *Leading American Inventors*, New York, Henry Holt, 1912.

Jennett S, *Pioneers in Printing*, London, Routledge & Kegan Paul Limited, 1958.

*Journal of the Franklin Institute*, Vol CXXIX, January 1890.

Legros L. A. & Grant J.C, *Typographical Printing-Surfaces*, London, Longmans, Green and Co, 1916.

Lehmann-Haupt H, Wroth L. C, Rollo G. S, *The Book in America*, New York, R. R. Bowker Company, second edition, 1952.

Leng Sir W, 'How we publish our papers', *Sheffield Telegraph*, 1892.

Letter to the Editor from Hugo Dalsheimer: 'Mergenthaler Reminiscence', *The Sun*, 17 December 1974, Baltimore.

Levine I. E, *Ottmar Mergenthaler – Miracle Man of Printing*, New York, Julian Messner Inc, 1963.

*Linotype News*, August 1936, Brooklyn. [Reminiscences of early Linotypes.]

Mengel W, *Ottmar Mergenthaler and the Printing Revolution*, Brooklyn, NY, Mergenthaler Linotype Co, 1954.

Moran J, *The Composition of Reading Matter*, London, Wace & Co Ltd, 1965.

Mott F. L, *American Journalism*, New York, Macmillan, third edition, 1962.

*New Wings for Intelligence*, Baltimore, MD, Schneidereith & Sons, 1954. A privately printed tribute to Ottmar Mergenthaler with an introduction by his son, Herman Mergenthaler.

Pearson F, 'Ottmar Mergenthaler', *The Business Printer*, Mergenthaler Number, July and August, 1935, Salt Lake City, Porte Publishing Company.

*Printing Metals*, Fry's Metal Foundries Ltd, Tandem Works, London, SW 19, revised edition 1966.

Roberts C. M, *The Washington Post – the first 100 years*, Boston, Houghton Mifflin Co, 1977.

Romano F, *Machine Writing and Typesetting (Story of Scholes and Mergenthaler)*, Salem, NH, GAMA, 1986.

Schlesinger C, (Editor), *The Biography of Ottmar Mergenthaler Inventor of the Linotype*, New Castle, DE, Oak Knoll Books, 1989.

Schlotke O, *Ottmar Mergenthaler's Jugendjahre*, Berlin, Mergenthaler Setzmaschinen-Fabrik GMBH, 1924.

*Scientific American*, 'A Machine to Supersede Typesetting', 9 March 1889, p 1.

Southward J, *Machines for Composing Letterpress Printing Surfaces*, 20 December 1895, Journal of the Society of Arts, pp 74–82.

Southward J, 'The Evolution of the Composing Machine', *Sell's Dictionary of the World's Press*, 1895. [Part of unnumbered supplement.]

Southward J, *Type-Composing Machines of the Past, Present and the Future*, A paper read before the Balloon Society of Great Britain, London, Truslove and Shirley, 1890.

Southward J. and Ross H. M, *Typography*, Encyclopaedia Britannica, eleventh edition, vol 27, 1910–11.

Special Newsletter 14 of The American Typesetting Fellowship, 1890: *The Pivotal Year In Man's Quest To Automate Type Composition*, 1990 Conference, Nevada City, CA.

Stonhill W. J, 'The Linotype Composing Machine Examined', *Sell's Dictionary of the World's Press*, 1895. [Part of unnumbered supplement.]

Swanberg W. A, *Pulitzer*, New York, Charles Scribner's Sons, 1967.

The History of *The Times*, 'The Thunderer' in the making 1785–1841, London, *The Times*, 1935.

The Linotype Company's works, Broadheath, Souvenir of the Inauguration, 14 July 1899, The Linotype Company Limited.

The Linotype: its history, construction and operation, London, The Linotype Company, 1st edition, 8vo, 1893.

'The Linotype Works at Broadheath', Supplement to *Linotype Notes*, July 1899.

Thompson J. S, *Mechanism of the Linotype*, Chicago, The Inland Printer Company, 1902; reprinted together with *The History of Composing Machines* in the same volume, but retaining the original page numbers, by Garland Publishing Co, New York, 1980.

Thompson J. S, *The History of Composing Machines*, Chicago, The Inland Printer Company, 1904.

Tracy W, *Letters of Credit*, London, Gordon Fraser, 1986.
Wallis L. W, *A Concise Chronology of Typesetting Developments* 1886–1986, London, The Wynkyn de Worde Society in association with Lund Humphries, 1988.
Williams E, 'The Linotype', *Inland Printer*, Chicago, October 1890–September 1891, vol VIII, pp 790–1.
Woolley E. M, 'Forty Years for Forty Millions', *System – The Magazine for Business*, USA, September 1908, pp 212–223.

## *Documents loaned by Miss Nancy Perkins:*

– Assorted letters from J. O. Clephane and L. G. Hine;
– Entries from Mergenthaler's diaries;
– Script of 'The Cavalcade of America' radio play, *Ottmar Mergenthaler and the Invention of the Linotype*; first broadcast at 8.00–8.30 pm on Wednesday, 1 December 1937, and repeated 12.00 midnight to 12.30 am the same evening;
– Letter dated 10 December 1938 written by Harry G. Leland of the Fred Medart Manufacturing Co to Joseph T. Mackey, President of the Mergenthaler Linotype Company.
– Typewritten translation into English of a German document about Mergenthaler's antecedents. [Claimed to have been written by a Nazi official to establish that his family had no Jewish connections.]

## *Whittaker H. P, Papers:*

– Notes about refurbishment of original Blower Linotype.
– Reminiscences of Matthew Whittaker.

## *Trade Journals:*

British printing trade journals examined for contemporary and historical reports:
– *British and Colonial Printer and Stationer*;
– *British Printer*;
– *L&M News*;
– *Linotype Matrix*;
– *Linotype Notes*;
– *London Typographical Journal*;
– *Paper Record*;
– *Paper & Printing Trades Journal*;
– *Press News*;
– *Printers' Register*;

- *Printing Machinery Record*, formerly the *Printing Machinery Register*;
- *Printing News*;
- *Printing World*;
- *Sales and Wants Advertiser*;
- *Scottish Typographical Circular*;
- *Stationer, Printer and Fancy Trades Register*;
- *Stationery Trades Journal*;
- *Stationery World & Fancy Trades Review*;
- *Typographical Circular*;
- *Vigilance Gazette*.

## *The popular press:*

British newspapers and periodicals examined for early reviews of the Linotype and advertisements to launch the Linotype Company – numerous publications including:
- *Life*;
- *City Press*;
- *The Financial Times* – most frequent advertisements, all full-page;
- *Hawk*;
- *The Manchester Guardian*;
- *Pall Mall Gazette*;
- *Railway Herald* and *Railway Press* – both published by Joseph Lawrence;
- *St James Gazette*;
- *Star*;
- *Sun*;
- *The Times*.

Relevant items from US newspapers and periodicals cited in the text.

# *Index*

Advertising campaign for the British Linotype Company 156–63
air-blast *see* blower
Altrincham [British Linotype plant] 151
*American Bookmaker* 155
American Newspaper Publishers Association [ANPA] 91, 145
American syndicate *see* New York syndicate
Anna [Ottmar's sister-in-law] 107, 119, 129
Antwerp gold medal, 1894 132, 134
Armstrong, Mrs Nancy [US Ambassador] 214
Arthur, Chester [United States President 1881–5] 22
Ashley, the Hon Lionel [Linotype Company director] 164–5, 172–3
assembler box [and auxiliary shelf] 186–7
attack on Linotype operator 204
automatic justification 30, 97, 100–5, 117, 160, 179–80
automatic metal feed 116, 188

Baltimore 5, 12, 18, 32, 44–7, 59, 65–6, 79, 83, 109–10
  Bank Lane shop *see* Mergenthaler at Bank Lane
  Camden Street works 22, 29, 47, 56, 58, 64
  Clagett and Allen streets 61, 63
  Locust Point, Baltimore 5, 61
  Loudon Park Cemetery, Baltimore 137
  Mergenthaler Hall, Johns Hopkins University 141–2
  Mergenthaler School of Printing 140–1
  Mergenthaler Vocational Technical High School 141, 144
  Preston Street works 56
*Baltimore Sun* 120, 136

band(s) 18–22, 178–9
banquet and demonstration of second band machine 22
base, star or claw *see* star base
Benton-Waldo engraver 48, 52, 60
Berger, Richard [Early colleague] 49–50, 53, 58–60, 122, 205
Bietigheim 10, 121
Biography of Ottmar Mergenthaler 5–6, 16, 19–22, 24, 27, 29, 35, 40, 68, 75, 87, 106, 110, 126, 135
Black, Sumter [Draughtsman] 35
Blades, William [British master printer] 194, 227
Blower Linotype 33–4, 45, 68, 79, 152–3, 175–6, 179–83, 186, 193–7, 202–5
Blower Linotype operators
  Brookes, C. W. 204–5
  Chant, Phil A. 204
  Good, Martin Q. 52, 202
  Leland, Harry G. [First Linotype operator] 19, 123, 154, 205
  Leonard, E. G. 204
  Machen, W. B. 153–4, 175, 195–6, 204
  Saunders, Harry P. 202
Board of the American companies 32–3, 36, 44–5, 47, 49–50, 54, 56–7, 59–62, 65, 68–72, 74, 77, 79, 83–5, 92–5, 99, 110, 113–5, 152
Bright, Jacob 79, 152–4, 161–9, 173–4, 227
*British and Colonial Printer and Stationer* 170, 173, 198
British Linotype Company *see* Linotype Company [British]
British Linotype syndicate 69, 77, 151, 153–4, 156, 194–5

British manufacturing rights 77
*Brooklyn Standard Union* 81
Bullen, Henry Lewis [Printing historian] 19, 58, 124, 152, 207
Burr typesetting machines 29, 37, 40, 48

Cam set [brain of the Linotype] 180, 184
Camp, Miss J. Julia [Typist/operator] 19, 21, 23–4, 71, 161, 195, 196
cash flow problems 60, 92–5
casting lines of type 3, 16, 19, 21–4
'Castle on the Rhine' 29, 62, 98, 132
Chamberlain Hotel, Washington, DC 22–3, 31
Chicago 15, 32, 107–8, 138
*Chicago News* 29, 56, 194–5
Church, Dr William [Pioneer inventor of typesetting machinery] 1
Clemens, Samuel 32
Clephane, James Ogilvie 2–4, 12, 14–9, 24, 29–30, 37, 50, 57, 65–6, 68, 75, 95, 97, 99, 119, 132, 136–8, 149, 160, 177, 209, 231–4, 227
   correspondence with Mergenthaler 106, 108, 110, 114–8, 132
constraints on materials 188–90
constraints on type design 190
Cooper Union of New York 132
Cottam, John Charles [Company promoter] 154–6, 162–3, 165–9
cycle of operations 178

Dalsheimer, Simon [Baltimore printer] 135, 140, 149
Davids, Charles H. 59–61, 64–5, 68, 70, 72, 75, 121–2
demonstrations in the UK 153–4
distribution
   of Linotype matrices 177–80, 182–4, 186
   of movable type 1–2, 14
Dodge, Philip Tell 31, 48, 60, 69, 89, 93, 96, 97–100, 105–17, 122–5, 129, 131, 137–8, 150, 186, 188, 227–8
double wedge spacer 92–3, 97, 101, 104
drop letter *see* two-line letter

Easton, Samuel Fyfe [Linotype Company secretary] 162–3, 166

Economic Printing and Publishing Company 175, 196, 204
Edison, Thomas Alva 59, 117, 123, 141
ejection of Linotype slugs 177–8, 180, 182, 184, 188
Electric Typographic Company of New York, NY 30, 97
electrotype(s) 16, 48
Elliott Cresson gold medal 132
Ensingen [where Ottmar repaired the village clock] 6, 9, 121
etaoin/shrdlu 45, 180, 183
experimental output 36
extraneous sorts *see* special characters

First band machine 18–9, 124
first elevator 184
*Fourth Estate* 127, 139, 149
Franklin Institute of Philadelphia 132
Friedenwald Company 135
Friez, Julien P. [Early colleague] 58–9, 205
Fuller, Samuel [Film maker] 211–2

Gally, Merritt [Printing machinery inventor] 21, 101
Germany 90, 108, 141, 145
Gilmore, John Farquar [Cottam's partner] 162, 164, 166
Girod, Ernest [Baltimore machinist] 149, 204
Gladstone, William Ewart [British prime minister] 154, 160
Goble, Corban [Professor and printing historian] 6, 23, 142
Greenleaf, Abner 19, 22, 59, 68, 71, 74–5, 115–6, 127, 131, 135–7, 183
Gutenberg [Inventor of movable metal type] 1, 140, 157

Hachtel 6–7, 9, 141, 143
   Mergenthaler museum 141
Hahl, August [Ottmar's step-cousin] 11–2, 14, 16–9, 23, 125, 147, 149
Hahl, Caroline [Ottmar's step-mother] 6, 68
Hahl, Louis [Ottmar's master] 10–1, 121
Hall of Fame [of Great Americans] 145
hand engraved masters 48

# INDEX

Harmsworth, Alfred [Newspaper proprietor] *see* Northcliffe, Lord
*Hawk* 156, 166–9
Hine, Lemon G. 18–9, 22, 24, 27, 30–4, 45, 47, 57–8, 66, 70, 72, 74–7, 79–87, 90–6, 97–8, 106, 114, 120, 122, 135, 137, 205
   correspondence with Mergenthaler 74–7, 79–87, 90–6, 188
   correspondence with Reid 33, 38–9, 46, 50, 66
   insecurity of position 87, 89, 97
Horncastle, W. R. [British advertising agent] 169
hot metal 1, 124, 177–8, 185, 195
Hutchins, Lee [Attorney, son of Stilson] 77, 79, 114, 153, 164, 193
Hutchins, Stilson 19, 22, 27, 29–30, 56–7, 59, 60, 62, 66, 68–70, 74–5, 77, 95, 113–4, 149, 151–3, 160, 162–5, 193–4, 207, 228
   correspondence with Reid 27–8, 35–6, 38, 57–8
   office in London, England 153–4, 176

Iles, George [American biographer] 12, 121, 138
impression machine 16
*Inland Printer* 88, 124, 138, 193–5
installation of first machine at the *New York Tribune* 40–1

Jeremiad of complaints 49, 51, 132
John Scott medal 132
Johnson, E. Kurtz [Member of Washington group] 18–9, 50, 60
Johnson, Milton White [Baltimore clerk, brother of above] 45, 50, 56, 59–60, 122
Judge Alfred Coxe [Decided against Typograph] 98–9, 132
Judge Lacombe [Decided against Typograph] 90
Judge Marcus Acheson [Decided against Typograph] 99–100, 132
justification 1, 21

Keyboard 46, 179–80, 183
Koenig [Inventor of steam press] 1

*L&M News* 101, 204, 206–7
Lambert, E. [Reid's assistant] 45, 50, 66, 72
Lambert, Ernest Orger [Cottam's partner] 154, 156, 162, 164, 166, 168–9
Lawrence, Joseph 69, 77, 79, 92, 151–3, 161, 164, 169, 174, 176, 200, 228
*Leeds Mercury* 173, 175
Letsch, Charles W. [Baltimore machinist] 40, 45, 86, 122, 124, 202–3, 205
Liederkrantz 17, 119
Linotype 1, 21, 27, 40–3, 47, 52–3, 58–9, 77, 90, 151, 153, 156–63, 197
   advantages 157, 159
   metal 58, 189
   naming 194, 207–8
   slugs 41, 48, 58, 98, 177, 182
Linotype advertisements 88, 91, 156–63
Linotype Company [British] 77, 79, 151–76
   factory 172–3
   orders in hand 173
   preference shares 174
Linotype Company [German] 141, 143
Linotype libel case 166–9
Linotype magazine 45, 179, 183, 185
Linotype maintenance and operation [Chapter 8] 177–92
   lubrication 192
   metal pot 190–1
   metal temperature 190
   rejuvenating Linotype metal 191
   space bands 191–2
Linotype matrices 21, 26, 33, 37–8, 44, 47–51, 192
   double-letter 186, 190
   electrotyped 37, 189
   hardened brass 189
   individual 179
   made by casting 37–8, 189
   steel 112, 116
*Linotype Matrix* 207–9
*Linotype News* 40, 203, 209, 211
Linotype, Blower *see* Blower Linotype
Linotype, Model 1 185, 190
Linotype, Simplex 68, 185
Linotype, Square Base 66–8, 71, 80–1, 88, 123, 170, 175–6, 183–5
lithography 2, 4, 15

Logotype machine  125, 131
*Louisville Courier-Journal*  29, 80, 195

Machinery Trust Limited  213
Mackey, Joseph T. [President of Mergenthaler Linotype Co]  19, 205
Manchester, England  79, 173, 196
manufacturing rights, sale abroad  69, 77–9, 152
Marble, Manton [Sometime proprietor of *New York World*]  2
Mergenthaler, Fritz [Half-brother Friedrich]  68, 107, 123, 149, 229
  Ronald [Fritz's grandson]  123
Mergenthaler, Johann Georg [Ottmar's father]  6, 9, 10, 112, 121
Mergenthaler Linotype Company  19, 94–5, 100
  Brooklyn works  59, 65, 75, 80–7, 99, 109–10, 202
  dividend(s)  116
Mergenthaler, Ottmar  229
  and Whitelaw Reid  29, 34–8, 44–54, 56–7, 59, 61–2, 64–6, 68, 71–3
  appointed Consulting Engineer  75
  at Bank Lane shop  18–21, 23–4, 74, 178, 205
  at Camden Street shop  22–5, 26–7, 29
  at the Hahl shop  11–7, 209
  children  119–20
    Eugene George [Third son]  120, 141, 143
    Fritz Lillian [Eldest son]  120
    Herman Charles [Fourth son]  100, 119–20, 124–5, 143, 150
      George Ottmar [Herman's only son]  120
    Julius Ottmar [Second son]  65, 120
    Pauline Rosalie [Daughter]  119–20, 138–9, 141, 143
  complaints about Dodge  109–14, 127
  contradicts Reid in open meeting  51
  death, will and estate  136–8
  Deming, New Mexico and prairie fire  129–31
  diary entries  79, 82, 90–1, 95, 98–110, 112–4, 125–6, 128–9, 131
  Emma [Wife]  6, 17–8, 107, 119, 121, 123, 129, 136–9, 206, 209
  health  39, 66, 113, 117, 119–20, 123, 125–31, 149
  industrial relations  121–3
  opinions of his creativity  123–5
  payment for tools  54, 56, 61–2, 64–6, 71, 73, 74
  Phoenix gold mine  129
  recognition  132–5, 138–45
  resignation  48, 53–4, 56–7, 59, 61, 74
  royalties  22, 24–5, 34, 49, 50–1, 54, 56–7, 62, 71, 73–6, 82, 93–5, 98, 114, 128, 132, 137, 150
  rumours of madness  24, 206–7
  salary  22, 75, 93
  under Dodge [Late 1891 onward]  74–96
  under Hine [1889–91]  81–2, 94
  youth and apprenticeship  6–10
Mergenthaler postage stamps  144, 146–8
Mergenthaler Printing Company  33–6, 60–1, 64, 81, 162
  Executive Committee  51, 58, 62, 69–70
Miller, John T. [*Tribune* operator]  40–1, 48, 202, 203, 206, 211
Miller, Thomas [Brother of John]  36, 41
Mills, Darius Ogden [Banker]  72, 94–6, 99, 108, 110, 112, 114
  Ogden [His son]  94–6, 112, 114
Monoline  97, 105–8
Moore, Augustus Martin [Editor of the *Hawk*]  166–9
Moore, Charles T. [American inventor]  3–4, 12, 16, 229
*Morning Herald*  57, 59
mould  16, 18, 180, 186–7
  liners  187
  recessed  187
  universal  187
  wheel  180, 184, 188
movable type  1, 14, 21, 29, 37, 43, 48, 58
Muehleisen, Carl [Baltimore machinist and superintendent]  123, 127, 149

Nailing the myths  205–12
National Inventors Hall of Fame  145

National Machine Printing Company 16–8, 149
National Typographic Company 18, 19, 22–4, 29–30, 33, 60–2, 64, 116, 153, 161
Nelson, Stan 214–5
*New Orleans Times-Democrat* 83, 86–7
New York 14, 32, 40–1, 47–8, 58–9, 67, 81–2, 131
*New York Herald* 68, 116
*New York Sun* 23, 30, 32, 45
New York syndicate 27, 29–31, 33–6, 38, 45, 57, 60, 62–3, 72, 77, 151, 205, 229
  Haldeman, W. N. [*Louisville Courier-Journal*] 29, 51, 195
  Lawson, Victor [*Chicago News*] 29, 57
  Rand, W. H. of Rand, McNally & Co, Chicago 29, 71–2
  Smith, Henry [*Chicago Inter-Ocean*] 29
  Smith, William Henry [Associated Press] 29–33, 38–9, 56–7, 61–2, 64–5, 70, 87
  Stone, Melville [Joint owner of the *Chicago News*] 29–33, 36
*New York Tribune* 1, 26–7, 36–7, 40, 58, 77, 151, 153, 212
*New York World* 2, 32, 89, 116, 211
newspaper advertising campaign in the UK 153–63
Northcliffe, Lord [Alfred Harmsworth, newspaper proprietor] 201
Northwestern Type Foundry 48

Oldach and Mergenthaler 205

*Park Row*, [Samuel Fuller film] 145, 211–2
Place, John [British inventor] 170
Printers' Composing Machine Company of Philadelphia 23, 30
*Printers' Register* 23, 154, 170
*Providence Journal* 59, 80
Pulitzer, Joseph [Proprietor of the *World*] 32, 59, 88, 90, 211
punches, steel 16, 19, 45–6, 48

QWERTY keyboard 37, 90, 180

Radio broadcasts 145, 209–11
*Railway Herald* 153
*Railway Press* 193–4
Randall [Brooklyn superintendent] 110
Ray, D. B. [American inventor] 101
Reid, Whitelaw 26, 29, 32–3, 35–7, 39–41, 48, 50–3, 60, 64–5, 68, 72, 74–5, 77, 94, 96, 98–9, 122, 132, 135, 150, 160, 172, 183, 202, 205, 207, 209, 212, 229–30
  and Mergenthaler 29, 34–8, 44, 46, 48–54, 56, 61–2, 64–6
  composing room expenses at *New York Tribune* 52
  correspondence with Dodge 48, 60, 69
  correspondence with Hine 30, 32–3, 38–9, 45–6, 50, 66
  correspondence with Hutchins 27–8, 35–6, 38, 56–8, 61, 69
  correspondence with Mergenthaler 37–8, 52
  correspondence with Wm Henry Smith 30–3, 39, 56, 62
  correspondence with Stone 30, 36
  lampoons 54, 210
  resignation 72–3
  1888 report 51–3, 71, 172, 183
  1889 report 54, 68, 70–1, 183
Reilly, Lee [Linotype operator] 196–7
Remington 2–3, 90
Rice [Temporary factory superintendent] 72, 75
Rogers Typograph 87–91, 99–101, 108
Rogers, John R. [Inventor of the Typograph] 87, 90–1, 97, 99–100, 230
Rogers, Rob [Legal assistant to Dodge] 106, 131

Schlesinger, Carl [Printer and printing historian] 37, 145
Schneidereith, Charles W. [Baltimore printer] 12, 149
Schoenrich, Otto [Writer of the Mergenthaler biography] 5, 129, 131
Schuckers, Jacobs W. 30–1, 33, 92, 97–104, 113, 117, 124, 230–1
*Scientific American* 41–2, 75, 153, 180

*Scottish Leader* 173–4, 204
Scudder, Wilbur S. [Inventor of the Monoline] 75–6, 83, 97, 105, 106, 108–9, 231
second band machine 19–20, 26–7, 31, 101, 124, 178–9
second elevator 184
self spacing type 48
*Sheffield Telegraph* 175–6, 197–9, 204
Sholes [Inventor of the 'Remington' typewriter] 2
single matrix machine 27, 33
single wedge spacer 21, 101, 104, 107, 114–6
Smithsonian Institution 214–5
Southward, John [British printer and author] 153, 156, 170, 172, 194, 200, 231
space band(s) 98–102, 180–2, 184, 191–2
space box 180, 182, 184
special characters 179–80
star base 185
Statue of Liberty 5, 26, 211
steel punches *see* punches, steel
step justification *see* single wedge spacer
stereotyping 16, 18, 21
*Sun* [British Sunday newspaper] 154–6, 166

Tabular work, tabulation 26, 52, 178–9
*The Daily Telegraph* [London daily newspaper] 214
*The Financial Times* [London daily newspaper] 157–63
*The Times* [London daily newspaper] 1, 2, 144, 156
Thompson, W. P. [*Tribune* machinist] 36, 45, 205
Thorne typesetter 170–1
transfer machine 4, 15
*Tribune Book of Open Air Sports* 40–1, 43, 124
Twain, Mark *see* Clemens, Samuel
two-line letter 113, 186, 214
typewriter(s) 2–3, 14

Unions 52, 68, 72, 81, 160
London Society of Compositors 160
Typographical Association [British printers' union] 160
Typographical Union [American printers' union] 68, 83
United States Patent Office 24, 92, 99–100, 106, 108
United States Postal Service 145

Walters, John [Former proprietor of *The Times* of London] 1
Washington Group of investors 19, 34, 60–1, 66, 70, 73, 137, 205, 231
Bryan, Samuel 94–5, 113
Clephane, Lewis [Brother of James O.] 12, 16
Crossman, J. H. 12
Devine, Andrew 75, 114, 150
Hume, Frank 18–9
Murphy, E. V. 114, 135
Pechin, Maurice 12, 16
Royce 3, 16
Warburton, Frederick J. 73, 75, 95, 137, 149
*Washington Post* 19, 28–9, 62, 68
Washington, DC 11–2, 15, 17–9, 22–4, 32, 56, 108, 128
Government Printing Office 28
Whittaker, Matthew [British engineer] 113, 152–3, 174, 186, 213–4, 231
Whittaker, Peter [Refurbisher of the Smithsonian Blower Linotype] 174, 213–7, 231
Wich, Ferdinand J. [Colleague of Mergenthaler] 19, 21, 149, 205
Williams, Eb [American Blower operator] 194–5
Williams, Chas R. ['Buyer' of the Baltimore plant] 65–6
Wright, Charles A. [Author of winning radio documentary] 209
Württemberg 5, 6